AMERICAN SCHOOL OF PREHISTORIC RESEARCH

PEABODY MUSEUM OF ARCHAEOLOGY AND ETHNOLOGY
HARVARD UNIVERSITY
BULLETIN 41

THE DISCOVERY OF GLASS

Experiments in the Smelting of Rich, Dry Silver Ores, and the Reproduction of Bronze Age-type Cobalt Blue Glass as a Slag

JOHN E. DAYTON

With an introduction by

C. C. Lamberg-Karlovsky
Peabody Museum of Archaeology and Ethnology

PEABODY MUSEUM OF ARCHAEOLOGY AND ETHNOLOGY
HARVARD UNIVERSITY, CAMBRIDGE, MASSACHUSETTS
1993
Distributed by Harvard University Press

Copyright © 1993 by the
President and Fellows of Harvard College
All photographs and illustrations
courtesy of John E. Dayton
ISBN 0-87365-544-3
Library of Congress Catalog Card Number 90-62664
Printed in the United States of America

Contents

Introduction

While John Dayton is by training a civil engineer, he is also somewhat of a polymath. For instance, he has constructed bridges; he has received significant prizes for flowers grown in his garden; he curates his own collection of minerals, gathered by him all over the globe; and he collects splendid Orientalist masterworks. His interest in architecture and interior decoration led him to renovate his London apartment in the eighteenth-century style of William Kent and Robert Adam, using Adam's 200-year-old molds for the plasterwork and period wallpaper made specially in Switzerland.

In the early 1960s, Mr. Dayton was commissioned to rebuild the Hejaz Railroad in Saudi Arabia, the very railroad blown up by T. E. Lawrence during the Great War. The railroad was not to be completed, but John Dayton convinced two archaeologists, Peter Parr of the Institute of Archaeology, University of London, and G. Lancaster Harding, then Director of Archaeology of the Kingdom of Jordan, to undertake an archaeological survey within the vicinity of the railway, the first by professional archaeologists in the Kingdom of Saudi Arabia. Decades were to pass before King Feisal promulgated an antiquities law and archaeologists were given a limited license to work in the Kingdom of Saudi Arabia. The paper published by Parr, Harding, and Dayton (1972) remains today one of the significant contributions to the archaeology of that country.

In retrospect, it may be said, to the dismay of some professional archaeologists and with the tolerance of others, that this Arabian experience led Dayton to become a practitioner in the field of archaeology and to spend four years under Professor Seton Lloyd at the Institute of Archaeology, London, completing the Diploma in Middle Eastern Archaeology. He then spent a further six years as a student of Henry Hodges working on a Ph.D. on ancient glazes. Whatever one thinks of as the typical archaeologist, and stereotypes abound, it can clearly be stated that he does not conform to any caricature. He has been an outsider, and outsiders are not readily tolerated by the card-carrying professional. Even though outsiders such as Willard Libby, Michael Ventris, Malcolm Wiener, and Thomas Jefferson have contributed significantly to the discipline of archaeology—as well as the late Marny Golding within Saudi Arabia—their practice is typically met with ambivalence from their professional counterparts. Why this should be so would require a contextual study of the sociology of the discipline.

John Dayton, while an outsider, has been involved with archaeology over the past three decades with a vengeance. His outsider status remains unchanged. As the founder of the journal, *Seminar for Arabian Studies*, which publishes the papers delivered at the Society for Arabian Studies, he provided the first professional context for scholars involved with the history and archaeology of the Arabian peninsula. He has been the secretary of that organization since its inception.

In this publication, John Dayton attempts to demonstrate through metallurgical experiment that some technologies long believed to have originated in the ancient Near East and then to have been transferred to Europe were, in fact, discovered and developed in Europe and moved eastward. One key artifact is cobalt blue glass found throughout the Levant and Egypt and in the graves at Mycenae. In its raw form, Dayton believes this glass could have come only from the rare native silver (of which it is a by-product) found in what is now the Moravian region of Germany. He further believes on geological grounds that tin bronzes and silver itself also originated in Europe and then were widely dispersed through trade to the East in the form of ingots ready to be locally processed.

It is important to point out that the experimental work reported here was done almost ten years ago. A cobalt blue glass was produced as a by-product of the smelting of a rare ore from the Erzgebirge region of Saxony and was undertaken prior to the discovery of the mid-second millennium B.C. Ulu Burun shipwreck off the coast of Turkey near Kaş, recently excavated by George Bass (1986). From that shipwreck more than 20 round glass ingots of cobalt blue glass were recovered. Irrespective of interpretation, chronology, and his reconstructed cultural processes, Mr. Dayton provides the reader with an outstanding example of experimental archaeology. With considerable prescience he addresses the origin of the blue glass ingots on the Ulu Burun shipwreck. His laboratory experiments with the silver ore from the Erzgebirge, and the cobalt blue by-product which he discovered in the smelt, have wide-ranging implications. His interest in minerals is deeply embedded in geological knowledge and in knowing what happens when specific minerals are subjected to pyrotechnological techniques. On his visit to the Erzgebirge, it was quite natural for him to collect samples and subsequently experiment with them. What was less expected was that in the process of the smelt he would discover a layer of cobalt blue glass resting atop the crucible. He immediately recognized its important implications.

In this monograph, John Dayton returns to many of his long-standing interests. Chief among them is lead isotope analyses, attested to by his publications in the bibliography. Equally clear is that archaeological artifacts, particularly minerals, have a geological context, and the geology must be understood. His frequent criticisms of archaeologists who undertake materials science approaches in the absence of geological knowledge has an undoubted ring of truth.

While geology and experimental approaches in the laboratory are fundamental to his eclectic approach, his reliance on the early literature detailing techniques of pyrotechnology as in the sixteenth-century writings of Agricola and Biringuccio are also fully in evidence. In tracing the history of cobalt blue glass, he notes that the Augsberg, Le Mans, and Canterbury cathedrals have windows of cobalt blue and that it was well understood that "colours which artists use" (Agricola 1530) were often derived from the slag by-product of silver smelting. His interest in paintings is no less in evidence in his casual aside that Leonardo da Vinci's "The Virgin on the Rocks" relied upon cobalt blue for its pigmentation.

Mr. Dayton makes a thoroughly admirable attempt to integrate historical, archaeological, geological, and analytical laboratory results into a seamless pattern. It stands in contrast to the professional who concentrates exclusively on the narrower confines of a particular specialization. The integrative approach is to be applauded *even* if weaknesses are apparent in specific aspects of his interdiscipli-

nary undertaking. These weaknesses and errors will be corrected in the cumulative advance of each discipline. An interdisciplinary approach is preferable to that of many archaeometrists who provide merely analytical results in the absence of understanding the historical, archaeological, or geological context of the problem their laboratory science is addressing. Needless to say, the person who attempts to "do it all" is vulnerable, for the chances of committing an error greatly increase. The risk of the undertaking, so frequently not tolerated by the professionals who find their territory invaded, is often worth the reward.

Mr. Dayton would readily acknowledge that in several areas his views of the past are positively idiosyncratic when compared to the perspective of most scholars. This is perhaps most evident in the chronological reconstructions which he put forward in his extensively documented volume, *Minerals, Metals, Glazing, and Man* (1978). His idiosyncratic, chronological world view of the second millennium B.C. continues to be reiterated. In a recent essay on faience (1989), he draws parallels between this material in the Aegean and the Indus Valley. The similarities in faience type and in its technology of production lead him to date the Indus Valley civilization to the middle of the second millennium. Quite simply stated, this is not a credible argument. Similar objects, indeed objects produced in an identical manner, need not be contemporaneous. The evidence for similarities in faience from the Aegean and the Indus Valley does not provide sufficient argument to place the Indus Valley civilization 500 years later than is generally accepted.

John Dayton, however, has a particular view as to the nature, extent, and timing involved in the transfer of technology in the ancient world, and he puts it forward in this monograph. He does not accept the discrepancy of well over 200 years between the dating of the earliest cobalt blue glass from Nitra in Slovakia, and in the Shaft Graves at Mycenae in Greece. "This is rather too long a time after the Nitra specimens to be credible, for *the spread of the manufacture of pretty blue glass would have been rapid after its initial discovery*" [emphasis mine].

John Dayton believes that technology transfer in the ancient world was a fairly rapid process. When an innovation or discovery appears in one place, it will spread *rapidly* to distant areas and afford a degree of chronological contemporaneity between the two places. In this regard he requires a chronological near-equivalence as an essential element in the presence of similar technologies in different geographical areas. This view tends to go against the evidence of the archaeological record. Technology transfer in the ancient world was a slow and seemingly stochastic process. It took thousands of years for an agricultural technology to reach northern Europe from the Near East. Metallurgical technology can be documented toward the end of the eighth millennium at Cayonu Tepesi in Turkey, where native copper was hammered into decorative items and small tools: fishhooks, pins, and awls. Yet the majority of sites in the ancient Near East dated to a thousand years later still lacked metal tools. The same can be asserted for the slow spread of iron technology. Rarely present toward the end of the third millennium, iron technology becomes common only at the end of the second millennium. And what of writing, a highly significant technology, which was simply not adopted by near neighbors and had a notoriously slow spread over the ancient world?

The rapid pace of technology transfer would seem to be a modern phenomenon. Recently, Philip Kohl (1989) has argued that in the ancient Near East technology was open-ended and readily transferred over considerable distances and between different cultures. This hypothesis allows for the free and ready transport of both technological skills and innovations in the absence of social and/or economic barriers. This assumption is presented by Kohl in the absence of evidence. The view that the past behaved as the present and that technologies spread rapidly is unfounded. Similarly, the view that technological secrets, so much a part of the present, were *not* part of a romantically reconstructed past may be equally erroneous. One of the most common colophons in the technological texts of the Babylonian period seems to indicate a tendency to keep technological secrets: "Let the initiate show the initiate, the noninitiate shall not see it. It belongs to the tabooed things of the great gods" (Saggs 1962:471).

In Mesopotamia the occupations of craftsman and scribe afforded social mobility. It is not unlikely that technological secrets would enhance one's position if not one's wealth.

Dayton's view that not all technologies represented in Europe were derived from the Near East is a healthy antidote to the years of *ex oriente lux* that dominated the field for generations. That the Egyptians and the seafaring Mycenaeans derived their cobalt blue glass from central Europe, specifically the source that Mr. Dayton cites, is a

clear advance on our prior understanding. But what are we to make of the fact that Robert Brill's analysis (cited in Bass 1986) indicates that the blue glass from Mesopotamia and Iran is of a different chemical composition? Analyses would seem to indicate that different sources were utilized in the eastern Mediterranean and Mesopotamia and Iran. If so, the sources for Mesopotamian and Iranian blue glass remain to be discovered. Perhaps we are confronted by a situation similar to the enigmatic problem concerning the location(s) of the tin sources utilized for the production of bronzes in the ancient Near East.

John Dayton has long grappled with this problem. He continues to favor a European source for all Near Eastern tin, denying the existence of an Afghan source (Cleuziou and Berthoud 1982), and ignoring the recent report of the discovery of tin sources in Anatolia (Yener and Uzbal 1987). It is clear that the last word on tin sources has not been written, but it remains entirely possible that both European and Asian sources were utilized. The search for distant tin sources continues and, as with lapis lazuli, these rare resources remind us that the peoples of the eastern Mediterranean, from the middle of the third millennium, were involved in long-distance trade that came close to being intercontinental in scale. Not so long ago, lapis lazuli was believed to be restricted to a single source, the Badakhshan Mountains of Afghanistan. More recently, with the excavations of sixth-millennium Mehrgarh in Pakistan and the recovery of lapis lazuli from that early context, the excavator Jean-François Jarrige has personally communicated to me the presence of another source of lapis lazuli in the Chagai Hills of Pakistan. Perhaps, in like manner, a silver source of comparable nature to that of the Erzgebirge will be located which provided Mesopotamia and Iran with its cobalt blue glass. If so, the archaeologist and historian of technology would confront the need to distinguish the independent invention of the production of cobalt blue glass within different geographic regions from the exigencies of technological transfer.

John Dayton's experimental laboratory work that produced the cobalt glass from the ores of the Erzgebirge deserves to be replicated. In it we have a contribution of the utmost significance. The Ulu Burun shipwreck offers a panorama of goods that were available to the eastern Mediterranean markets about 1450 B.C.: ivory, copper and tin ingots, faience beads, rhyta, and cobalt blue glass ingots. Regarding the latter, Dayton is explicit: "The cobalt glass ingots could only have come from the Erzgebirge." Thus, it is the only commodity on shipboard which has a definite point of origin. The geographical origin for the ivory, tin, copper, and faience remains the source of great speculation, as does the point of origin, destination, and ethnic origin of the ship's crew. Mr. Dayton, never hesitant to offer his views, has an interpretation on all of the above. True to form, he does not disappoint: his views contest those of most establishment scholars. It is those same scholars who will view this monograph with raised eyebrow. In being too

willing to negate his interpretations, they will miss his substantive contribution: a remarkable example of experimental archaeology which leads to an understanding of both the technology of production and the geographical provenience for the manufacture of cobalt blue glass.

The experimental laboratory work that this volume reports was undertaken almost a decade ago; the manuscript was completed five years ago, and the material on Ulu Burun incorporated more recently. The Peabody Museum is pleased to present this provocative monograph.

C. C. Lamberg-Karlovsky, Director
June 1990

Acknowledgments

With the very kind help of Herr Siegfried Flach, the expert on the Erzgebirge region of Germany, I obtained a rare sample of cobalt blue glass from Schneeberg itself (pl. 3e), together with an equally rare sample of lead from this normally lead-free deposit. I also wish to acknowledge the great assistance given by Jenny Inge and Dick Springer of Colorado; by the late Jack Still of Tucson, who kindly obtained a sample of ore from the Monte Christo Mine at Wickenburg; by Charles R. Sewell, also of Tucson, for help with Mexican samples; by Winthrop R. Rowe III; by William J. Robertson and Edward M. Tomany of Tonopah; and by numerous other friends of the Mining Club of the Southwest, Tucson, Arizona.

Thanks are also due to Dr. John Bowles and David Bland of the British Geological Survey, to Dr. Peter Emburey of the British Museum (Natural History), and to Dr. Valerie Chamberlain of the University of Alberta, for their analytical help and advice; to Professor C. C. Lamberg-Karlovsky, Director of the Peabody Museum, Harvard University, for kindly reading the manuscript and for much encouragement and advice; to Dr. G. Padolino of Cagliari for samples of cassiterite in quartz from the granites of Arbureze in southwest Sardinia, and of copper ores from the island; and to Beatriz Hernandez of La Laguna, Gisella Young, and Jamel and Nahid Burna Asefi for help with the maps of Spain.

Finally, thanks to Professor George Bass, Nieta Piercy, and the director of the Bodrum Underwater Archaeological Museum in Turkey for allowing me to make a very full and detailed study of the material from the exciting and important Kaş shipwreck with its circular ingots of cobalt glass. These confirmed my hypothesis that the discovery of glass was an accidental by-product of the smelting of a very rare silver/cobalt ore found in pure white quartz in central Europe.

MAP 1

The Mineral-Rich Iberian Peninsula

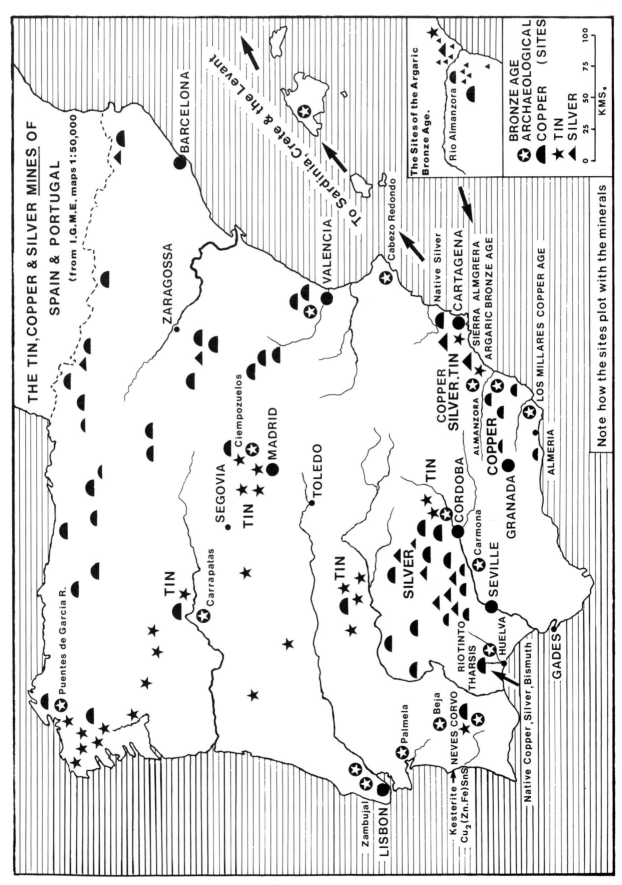

THE TIN, COPPER & SILVER MINES OF SPAIN & PORTUGAL
(from I.G.M.E. maps 1:50,000)

To Sardinia, Crete & the Levant

BRONZE AGE ARCHAEOLOGICAL (SITES

The Sites of the Argaric Bronze Age.

Rio Almanzora

- ✪ COPPER
- ◗ TIN
- ★ SILVER
- ▲ (legend symbol)

KMS. 0 25 50 75 100

Note how the sites plot with the minerals

BARCELONA

ZARAGOSSA

VALENCIA

Cabezo Redondo

Native Silver

CARTAGENA

SIERRA ALMGRERA

ARGARIC BRONZE AGE

LOS MILLARES COPPER AGE

COPPER
SILVER.TIN

ALMANZORA

ALMERIA

COPPER

GRANADA

CORDOBA

Carmona

SEVILLE

MADRID

Ciempozuelos

SEGOVIA

TIN

TOLEDO

TIN

TIN

SILVER

Carrapatas

TIN

Puentes de Garcia R.

RIOTINTO

THARSIS

HUELVA

GADES

Beja

NEVES CORVO

Palmela

← Kesterite
Cu₂(Zn,Fe)SnS

Native Copper, Silver, Bismuth

Zambujal

LISBON

$Cu_2(Zn,Fe)SnS$

1 EVIDENCE OF EARLY EUROPEAN GLASSMAKING AND METALLURGY

A thorough examination and analysis of archaeological artifacts and data, as well as metallurgical experimentation, would indicate that man's first use of advanced metallurgy, the deliberate use of tin with copper to make a bronze, developed in the metal-rich areas of central and western Europe many centuries before such technology reached the Fertile Crescent (Egypt, Syria, and Mesopotamia), up to now considered the "Cradle of Civilization." Carbon 14 dating demonstrates that the use of tin bronzes spread all over Europe at about the end of the third millennium B.C.: i.e., some time before 2000 B.C., hundreds of years before tin bronze arrived in the Aegean, the Levant, and the Middle East. Further, the Fertile Crescent had no deposits of the vital tin and copper necessary to make bronze.

On geological grounds, the tin must have come from central Europe, Sardinia, or Spain (if one rules out the rich tin deposits of Burma, Malaysia, and Indonesia). Silver also appeared at the same time and was not found in the Fertile Crescent. Cuneiform texts tell us that Sargon, ruler of Mesopotamia around 2300 B.C., had "a silver mountain that lay over the sea [Mediterranean] and to the west." In fact, apart from gold in Egypt, every metal object and semi-precious stone found in the civilizations of Egypt, the Levant, and Mesopotamia had to be imported. Even good timber for seagoing ships did not grow in Egypt or Mesopotamia.

Cobalt blue glass is also found with the bronze and silver artifacts in the Middle East. Cobalt, a silvery metal in appearance, is even scarcer than tin and only produces its beautiful blue color under certain conditions. This author has proved experimentally that cobalt blue glass could only have been a by-product of the smelting of a very rare, rich, silver ore found with native silver only in the Erzgebirge of Saxony and later in Ontario, Canada. Besides cobalt blue, the existence of white glass opacified with antimony, and yellow glass colored with lead and antimony is further evidence of man's breakthroughs in metallurgy and technology at the end of the third millenium B.C.

Lead isotope analysis confirms that many of the tin bronzes, silvers, and glasses of the Middle and Late Bronze Age in the eastern Aegean and Near East, about 1700 to 1200 B.C., derived from classic and historically known Sardinian, Spanish, and central European mining areas, and that the sulfide copper ores of Cyprus were first exploited by the Mycenaeans around 1400 B.C.

Up to now, archaeologists have thought that the early civilizations of Mesopotamia and Egypt were responsible for all the technological advances which were then diffused to the Levant, Minoan Crete, and Mycenae, and much later to barbarian Europe. It can now be shown that this diffusion was in the reverse direction from sophisticated metal-using peoples in the west to the east and that there were three routes:

1. From Spain via the Balearics, Sardinia, Tuscany, and Crete.
2. From the Únêtice Bronze Age culture of central Europe via northern Italy, the Adriatic, and early Mycenaean Greece.
3. From the Carpathians and the Balkans via the Danube and Anatolia.

It is clear that with the advent of metals, widespread Bronze Age trade routes existed, especially maritime routes, as the ships depicted in Cycladic frying-pan vessels, rock carvings, and the later Thera frescoes of about 1450 B.C. demonstrate. The lead isotope analyses confirm the existence of this trade over long distances.

Glass and Faience

It is a curious fact that the earliest glass known, the beads from Nitra on the river of the same name at the western edge of the Hungarian Ore Mountains, owes its blue color entirely to the scarce metal cobalt. Carbon 14 has revolutionized the dating of Europe, and Nitra belongs to the early Únêtice period (Coles and Harding 1979:49) (see map 2). During the Únêtice period the first true 10% tin bronzes appear. Ñivnáç has yielded a Carbon 14 date of 2434 B.C., and Nitra, dates of 2495, and 2310; Lobnitz, a Bell Beaker date of 2030, while ripe Únêtice dates are Prasklice 1895, Helmsdorf 1775 and 1663, Leki Mali 1655, and Leubingen 1675 B.C. The above dates are all uncorrected. This writer would put the appearance of true tin bronzes in the Únêtice culture at about 1800 B.C. or even later (see Dayton 1978:414), as, apart from the Nitra beads, the next cobalt blue glass beads appear in the Shaft Graves at Mycenae, around 1650 B.C. This is rather too long a time after the Nitra specimens to be credible, for the spread of the manufacture of pretty blue glass would have been rapid after its initial discovery. Europe experienced a warm period around 1800 B.C., when the Brenner Pass opened. This situation would have helped in the transmission of glass technology.

A feature of the Mycenaean world was the enormous quantity of cobalt blue glass produced (see pl. 6a,b). In excavations, it was discovered alongside a great wealth in metals, tin bronzes (90 bronze swords in Shaft Grave V), much silver and electrum, rock crystal from the Alps, amber from the Baltic, and amethyst found in quantity in the silver and tin mines of Saxony. The Mycenaean civilization was clearly in contact with metal-rich areas which enjoyed the availability of high technological skills. The cobalt glass, the tin, the amethyst, and the rock crystal all point to central Europe. Lead isotope analysis of the Haghia Triadha oxhide ingots (Dayton 1984), however, shows a thriving copper industry based on Sardinia, with contacts via the Balearics to the metal-rich regions of southern Spain (Dayton 1971). Central Europe and the western Mediterranean were far from being areas of darkness and ignorance.

The Prehistoric Museums of Hanover and Munich have enormous quantities of ingots, torques, and molds dated to around 1800–1500 B.C. from the Regensberg area; from Luitpold Park, Munich (see pl. 7); from Kelheim, Ljubljana, and Vucedol (2300–2100 B.C.); and from Margarethenberg (ca. 1600 B.C.). An interesting mold for 30 spherical beads comes from Nitriasky-Hradok and is dated at about 1500 B.C. The Munich Museum also has molds for typical Mycenaean ball-headed pins from the antimony-rich area of Velem St. Veit.

By 1400 B.C. the Mycenaean glass industry was at its peak. The glass beads and pendants (often very dark blue) were colored by cobalt, and often covered with gold foil: i.e., fake jewelry (pl. 6b). In Egypt, at El-Amarna around 1379–1362 B.C., cobalt was used extensively in faience and glass manufac-

ture, often with traces of copper. Earlier, around 1417 B.C., a cobalt blue pigment appeared on clay pots at Knossos on Crete, Tiryns in the Peloponnese, and in Egypt. Previously, in the Hyksos period around 1675–1567 B.C., silver first appeared in quantity in Egypt and the Near East.

After 1300 B.C. the supply of cobalt seems to have been disrupted, and there is a curious hiatus until glass appears again in large amounts with the Phoenicians, perhaps as late as 600 B.C. The body of this glass is once again colored by cobalt, often almost black in color (pl. 6c). At the same date, a flourishing glass industry existed in the Hallstatt culture of Bohemia and Austria (pl. 6e), and in northern Italy, where it is suggested that the Phoenicians were obtaining their raw material for glassmaking. From the Phoenician world polychrome glazing, again using cobalt among its colors, spreads to Babylon, Urartu, and to the Achaemenid world: e.g., to Susa.

Not long ago, glass and faience beads found in Europe were automatically assumed to be Egyptian imports. Stone and Thomas (1956) remarked on the high percentage of cobalt in two beads from Moravia, the homeland of the Únêtice culture, while earlier Lucas (1928, rev. 1962:189) concluded: "Egyptian glassmakers of this time [Amarna] were in contact with glassmakers elsewhere who were using cobalt." Sayre (1975) noted the similarities in composition of two cobalt blue Egyptian Eighteenth-Dynasty beads with two from Mycenae, while Harding and Warren (1973) found European faience beads to contain high amounts of antimony and cobalt. The writer found the cobalt blue beads of Hallstatt also to be high in antimony (Dayton 1978:423). Harding and Warren (1973:64) found central European beads to be high in silver, which the writer also found to be true of a dark blue frit from Amarna and a blue bead from the Mycenaean Menidi Tholos around 1400 B.C. (Dayton 1978:453).

Silvers from Ur showed traces of nickel, cobalt, antimony, arsenic, tin, and zinc, implying that the silvers had been melted and smelted from what are geologically known as the dry silver and lead-free ores. Silver is closely associated archaeologically with cobalt blue glass. In an important paper, Craddock (1976:93–113) found Middle Helladic coppers and bronzes to be often rich in bismuth, with traces of nickel, cobalt, and silver. Schliemann (1878) found similar traces in Aegean bronzes, as did Dörpfeld (1902:421) in Trojan bronzes. Otto and Witter (1952:138) analyzed 16 celts from Carlsdorf near the mining area of the Erzgebirge in Saxony which averaged 4% tin and 1% silver: all had traces of cobalt, nickel, and bismuth. Eighty-six ingot torques from the same area (ibid.) averaged 1% silver and considerable traces of bismuth. Silvers from the Mycenaean Shaft Graves analyzed by Filippakis and the writer (unpublished) contained significant traces of copper, cobalt, bismuth, and nickel. Gold was not present, and is significantly absent in the silver ores of the Erzgebirge.

All the above were important clues for the origin of cobalt blue glass, which after the Phoenicians next appeared in quantity in the great cathedrals of Europe in the twelfth century A.D., when the silver mines of the Erzgebirge were opened again.

During the great period of metallurgical innovation, some time after 1800 B.C. in central and western Europe, the lead-antimony ores of Bohemia and perhaps Tuscany were used to produce a yellow glass. In turn, the pure antimony ores of Pezinok and Velem St. Veit produced a white glass (see pl. 6c,d,e). It is of interest that three Beaker-type-V buttons of antimony and four daggers of Bohemian type were found at Monte Bradoni near Volterra, in the ore mountains of Tuscany (Trump 1966:69). Four "silver" beads from Tell Fara (Beth-Pelet), Tomb 226, analyzed by the writer, and of the Late Bronze Age around 1200 B.C., consisted of 66% tin and 33% antimony with no trace of silver. The tin and antimony must have come from central Europe where such mixed ores occur, as it is unlikely that tin and antimony were deliberately mixed.

The first yellow and white glass known to the writer appears in Egypt in the reign of Thothmes III, about 1450 B.C., or even a little earlier. A yellow Hyksos-type goblet is decorated with white antimony and cobalt blue (see Dayton 1978, pl. 12:7,10). A polychrome model of a glass coffin (British Museum Collection) of the same period has been analyzed and discussed (Shore and Bimson 1966:105). Some of the vessels carried by Keftiu in Eighteenth-Dynasty tomb scenes appear to be of such glass, as neither antimony nor cobalt existed in Egypt (see fig. 6). It is curious that glass, as distinct from faience, seems to die out after the Eighteenth Dynasty, only to reappear some 700 years later with the Phoenicians. Here, again, the main body of Phoenician glass vessels is of dark cobalt blue with yellow and white glass trailed in for decoration (pl. 6c,d). The same colors and techniques appear in the contemporary Hallstatt and later La Tène cultures with their superb "eye" beads and

arm rings (see pl. 6e). In the Levant, the Phoenicians did not have the coloring metals to produce such glass and must have traded with the glazing centers of Hallstatt and the Veneto at the head of the Adriatic, who in turn were trading with the cobalt- and antimony-rich areas of central Europe. Such contacts existed in the earlier northern Italian Copper Age (ca. 2500–1800 B.C.; Barfield 1971:56) when Sardinian obsidian was traded to northern Italy, and the ubiquitous Beaker Folk were present in the northern Italian Plain.

At the lake village of Lucone there are glass frit beads, while at Ledro there are beads of Baltic amber. The glass beads provide an important chronological link between those of Nitra about 1800 B.C. and those of the Shaft Graves of Mycenae about 1650 B.C. Furthermore, copper ring-ingots of Únětice type are found in the Adige Valley.

Barfield (ibid., 81) notes indirect evidence of contacts with Mycenaean Greece, such as a copy of a Vaphio-type cup made from local clay found at Ledro. Other curious finds, perhaps tallies used in trade, are little fired clay plaques, incised before firing with a variety of symbols and, according to Barfield, identical with similar plaques from Slovakian and Hungarian Bronze Age sites, and which he suggests are prototypes of the later Linear A tablets. For Levantine connections with Middle Bronze Age Europe, there are the rich finds from the Hyksos site of Gaza, with toggle pins in silver and gold, faience and dark blue glass, amethyst scarabs, weights, daggers, and long swords (see Dayton 1978, fig. 41; Petrie 1931; and Petrie 1934).

Gimbutas has written, "the central European Únětice-Tumulus-Urnfield sequence in Europe has never been described as a whole." She would put the beginning of Únětice at around 1800 B.C. and has an excellent map showing the distribution of faience beads at about 1400 B.C., beads which she thinks are of Aegean inspiration. Late Únětice Moravian beads have a dark, cobalt blue color, and she states correctly that glass beads of the early Urnfield period from Volders in Austria were produced locally (Gimbutas 1965:45).

Excavations at the large site of Frattesine, near Padua, have unearthed quantities of metal-working slag, as well as lumps of blue and red glass, the raw material for making beads (Coles and Harding 1979:49, 120, 417). This site is placed in the final period of the Late Bronze Age in northern Italy and unfortunately is not precisely dated.

It is now quite clear that all glass and faience beads did not come from Egypt. Faience or de-

FIGURE 1

The Great Bronze Sword

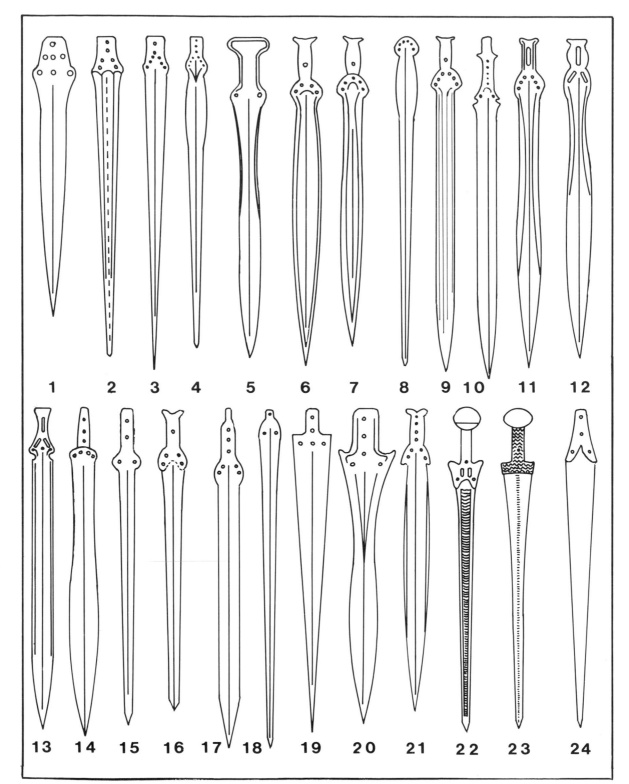

FIGURE 1

The Great Bronze Sword *continued*

The remarkable spread of the great bronze sword over western Europe and the Aegean, as far as the Levant and Egypt with the Sherden. Nos. 1–5 are from central Europe, near the tin and copper deposits of the Erzgebirge. Nos. 6–10, and 17 show the spread of the sword throughout Italy; 11–13, Spain; 14, Sardinia; 15–16, the Aegean. No. 24 is from Gaza, the large Hyksos *tell* which ended about 1481 B.C. and was contemporary with the Tholos Graves of Mycenae. Cobalt blue glass was found there together with a wide range of Mycenaean objects by Petrie (see Dayton 1978: fig. 41 for finds which included a haematite cylinder seal depicting Sherden-type swords). Nos. 18–23 are from the Shaft Graves of Mycenae, where cobalt blue glass appears for the first time in the Aegean (Schliemann 1878:152, 157–158). Three distinct types of sword are found in the same Shaft Graves IV and V; the Sherden type, nos. 19, 23; the rapier type, 18, 22 (without the hilt); and the type shown in nos. 5–7, 10, 12, 14 from central Europe, Italy, Sardinia, and Asturias Huelva, Spain.

No.	1	Bodrogkerestur
	2	Sombor
	3	Smolenice
	4	Keszthely II
	5	Nurnberg
	6	San Marco di Belvedere
	7	Montegiorgio
	8	Treviso
	9	Treviso
	10	Allerona
	11	Vila Maior, Sabugal
	12	Asturias
	13	Huelva
	14	Oroe, Siniscola, Sardinia
	15	Knossos
	16	Enkomi
	17	Lake Trasimeno, near Perugia, Italy
	18	Mycenae, Shaft Grave V
	19	Mycenae, Shaft Grave IV
	20	Mycenae, Shaft Grave IV
	21	Mycenae, Acropolis (Schliemann 1878:221)
	22	Mycenae, Shaft Grave IV, Karo Type B
	23	Gaza

vitrified glass beads have been found at Knossos, Malta, Fuente Alamo (in Cist Grave 9 with a sword and part of a silver diadem), El Argar in Spain, and in Wessex barrows together with Mycenaean-type pins (Turnham 1873:494; Evans 1921, fig. 228). Mycenaean sherds have been found in Italy (Trump 1980:192 ff.; Coles and Harding 1979:184, 440) and, importantly, in Sardinia (Ceruti 1979; Vagnetti 1980). In this latter location, another branch of metallurgy flourished, with connections to Corsica [where there is a deposit of copper, zinc, and *tin* 60 kilometers along the coast north of Ajaccio], the Balearics, and the rich metal deposits of southern Spain (see map 1). At a later period there is a remarkable quantity of glass of the Hallstatt and La Tène cultures that was reported by Hencken (1978) and Wells (1981:105, 113). Six graves of the Early Iron Age, excavated at Stiçna, produced some 20,500 glass beads (Haevernick 1974a; 1974b). Analyses of beads from the Mecklenberg Collection at the Peabody Museum [with the kind permission of Professor C. C. Lamberg-Karlovsky] showed that all four typical dark beads contained cobalt, copper, antimony, and tin.

Metals

In the third millennium B.C., the Megalith Builders spread all over the metal-rich areas of central and western Europe, but had no metals. Next, the Beaker Folk, dependent on bows and arrows, appeared with the late Megalith Builders. The Beaker Folk had simple copper daggers and, as Hencken pointed out long ago (1932), the Beaker Folk are found all over western Europe wherever there are metal deposits (see map 1). In the Polada culture of northern Italy, just north of the Brenner Pass, true tin brozes dated to around 1800–1450 B.C. were found. The presence of molds, crucibles, and clay tuyères used as bellows indicates that the people were advanced bronze workers. "The metal objects reflect a technology derived in its entirety from the Bronze Age cultures lying to the north and east of the Alps, notably Únêtice and Straubing, and owe nothing to inspiration from the Mediterranean world" (Barfield 1971:75).

With this technical revolution—the smelting of tin-bronze and its casting in bivalve molds—there occurred a revolutionary weapon of war, the great bronze sword. This sword was the distinctive artifact of the European and Aegean Bronze Age, whose leitmotif, according to Renfrew, was "warfare and wine" (see fig. 1). In eastern Europe the cast, sock-

eted battle axe was found in either the ingot form or as a weapon.

Bronze swords spread throughout Europe about 1700 B.C. or earlier. Their typology and distribution has been commented upon by many writers. Cowen (1955) described the enormous quantity of these swords in central Europe which spread along the axis from Vienna to Budapest, Klagenfurt, and throughout Italy (Bietti Sestieri 1973:405); as well as to Albania (Hammond 1967), Huelva in Spain (fig. 1:12,13) and the eastern Mediterranean (fig. 1:16,24). Catling (1964) postulated a definite route down the Adriatic for the Sprockhoff IIa-type sword (1964). He also described (1956:12) the Naue II swords from Khalandritsa, Enkomi (fig. 1:16), and Scutari (Catling 1956, pl. IX).

At a late phase of the Polada culture, at Bor di Pacengo on Lake Garda, and nearby Peschiera, were masses of bronzes, including swords, spearheads, and hundreds of diagnostic pins with all manner of heads (Barfield 1971:84; Dayton 1978, figs. 31–38), showing clear central European connections with northern Italy. The swords are particularly interesting when one considers those of the Shaft Graves of Mycenae, and the later Sherden-type sword of the Rixheim-Monza school, an example of which was found at Ugarit bearing the cartouche of the pharaoh Merneptah, around 1234–1220 B.C. Nor must one forget the enormous bronze hoards from the Bologna area, one with 14,838 scrap objects including 3,952 socketed axes.

Snodgrass (1974, pl. 33) illustrates a cut-and-thrust sword from the Siteia region and a Minoan rapier some 36 inches long from Arkalokhori in Crete. For an examination of the Carpathian material, see Mozsolics on swords (1973:258 (16), 266 (1–4), 274 (1–4), 275 (1), 323 (16)) and molds (ibid., 360–364). His analyses (ibid., 230–239) pinpoint considerable traces of antimony, nickel, bismuth, silver, and cobalt, as well as tin, but very little gold. The appearance of these trace elements show that the tin-rich bronzes were coming from the copper and the ores of the Erzgebirge, not from Hungary and the Balkans (Banat), where there is no tin.

Riem (1974) has discussed the finds of bronze swords in Germany, France, and Switzerland, as has Bianco Peroni (1970) in sites throughout Italy. The latter includes the discovery of three interesting molds for three types of swords at Pivarone near Turin (ibid., pl. 25:168–170) which were produced at the same center and the same time, demonstrating the hazards of overreliance on typology.

If the bronze wealth of central Europe is as enor-

MAP 2

Carbon 14 Dates (uncorrected)

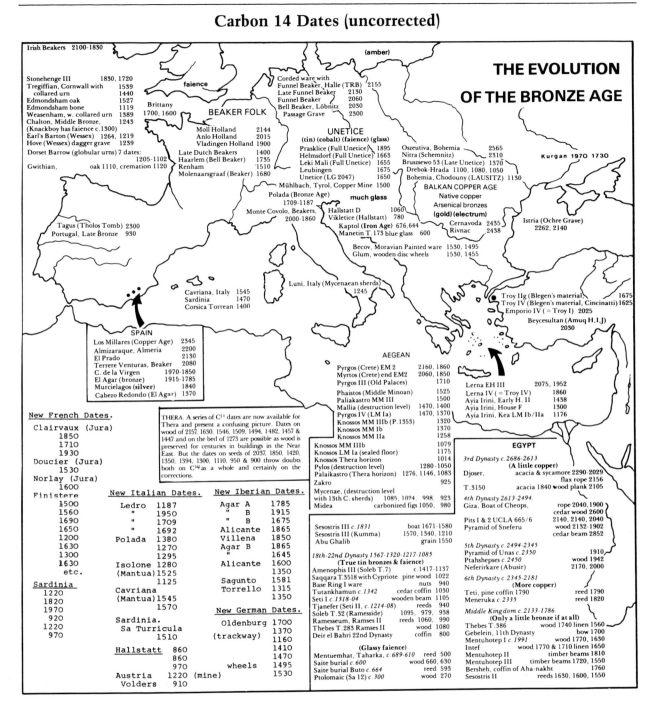

Carbon 14 has revolutionized the picture of European archaeology. The dates above relate to the appearance of metal artifacts: first, copper, then bronze. It is believed that the discovery of metallurgy took place all over Western Europe at and after 2000 B.C. The advent of metallurgy was much earlier in the metal-rich areas of Europe than in the Aegean and Near East. The new objects spread over this vast area from Spain to the Levant—swords, daggers, and cast, socketed axes—showed a remarkable uniformity, and replaced the bows and arrows of the Beaker Folk. The trade in bronze artifacts was widespread from metallurgical centers and the diffusion of technology to these centers, rapid.

mous as the collections in the Prehistoric Museum in Munich suggest (see pl. 7), so is that of northern Italy and the Shaft Graves of Mycenae, where one warrior was buried with no less that 90 swords. At Mycenae, there is even a sword hilt of cobalt blue glass that was once covered with gold foil (see pl. 6a) together with much cobalt glass.

The Shaft Graves of Mycenae, especially the very rich Graves IV and V, have three types of sword, *all contemporary* and showing affinities with central Europe, Italy, northern and southern Spain, Portugal, and Sardinia. The Sherden-type sword was present in the Shaft Graves of 1600 B.C. and continued to be found to the time of Ramses III, some 400 years later (fig. 1:19, 23)! The engraved ring from Shaft Grave III shows a warrior with this Sherden-type sword (Dayton 1978:252). It is clear that Mycenaean warriors went into battle with a rich assortment of weapons. Mylonas (1973:427) notes that their swords do not appear to have been remelted but buried with them as trophies. Both Schliemann's analyses (1878:157) and those of Mylonas (1973) showed the swords contained 13% tin with traces of antimony, nickel, and cobalt. Silver does not seem to have been sought. The graves were fabulously rich in silver, gold, electrum, rock crystal, Baltic amber, and amethyst. Lapis was not present.

In the contemporary Argaric Bronze Age culture of Almeria in Spain, women were buried with silver diadems and men also with long bronze swords. This shows the remarkable uniformity of the western European Bronze Age and the trade and cultural interchange that must have taken place over great distances. Also found in many parts of Spain and Portugal are the long-hilted, cut-and-thrust-type sword and, particularly, the remarkable Huelva hoard found at the mouth of the Rio Tinto, near to the copper deposits and not far from the rich tin deposits of Portugal (Savory 1968:232). Swords are depicted on rock surfaces and cist slabs from the Remedello and later Bronze Age cultures of northern Italy as well as in Corsica, Galicia, southern Portugal, and Britain. Savory (1968, fig. 70a) illustrates some of these carvings with their long swords, while at Conjo in Portugal, a Sherden/Philistine headdress was found (ibid., fig. 70c). Rock

carvings of swords in Corsica have been compared by Grosjean (1966:194–198) to those of the Sherden.

As long ago as 1887, the Siret brothers pointed out the parallels between the true Bronze Age cultures of southern Spain (El Argar) and the Polada culture in northern Italy, as well as the Únêtice culture in Bohemia. Many theories on the diffusion of copper and bronze technology have been proposed. That this diffusion took place quite rapidly over the whole area of western Europe and spread eastwards to the Aegean and the Near East cannot now be doubted. Harrison (1980) places the Argaric culture between about 1850–1700 B.C. and 1400 B.C. [This writer sees no reason for it not to continue to 1200 B.C. or even later, based on the evidence of the Sherden swords found among the Sea Peoples at the time of Ramses III.] Trade links from Almeria existed with the Balearics, where at the site of Son Matge there was a workshop which made tools of imported metal. There were crucibles, awls, clay molds, and incised pottery showing links with Sardinia and Sicily. Bray (1964) notes that the beakers in Sardinia were intrusive and must have come from a point further west, rather than Italy, and then spread by sea to Sicily.

The above would suggest that another look needs to be taken at previous assumptions with regard to the origins and dating of many important artifacts. Renfrew (1973) wrote that

> the study of prehistory today is in a state of crisis. Archaeologists . . . have realised that much of the prehistory as written in the existing textbooks is inadequate: some of it quite simply wrong. In Europe the conventional framework of our prehistoric past is collapsing about our ears. . . . There is a serious flaw in the archaeological theory in general.

Renfrew was, of course, referring to the arrival of Carbon 14 dating which showed that the megaliths of western Europe were considerably older than the monuments of the Near East.

The writer has tried to show in his book (1978) and here, that western Europe with its mineral deposits, especially those of copper, silver, and tin, was the home of the Bronze Age, and that metallurgy spread from west to east.

2   THE LOCATION AND AVAILABILITY OF METAL IN THE ANCIENT WORLD

Geologically there exists a scarce and distinct type of silver ore, the "silver-cobalt-nickel-arsenide" (Ag-Co-Ni-As) group, with which bismuth is nearly always associated (Stanton 1972:603–607) (see pl. 1b). The veins can be from two inches to five feet wide. A little copper is sometimes found but lead is normally absent, as is gold, indicating these ores as the source of the very pure silver of antiquity. In the Old World these deposits exist in the north-south fault at the western end of the Saxon Erzgebirge, where they cut the older east-west tin veins of Hercynian age (see fig. 9a). A small silver deposit is found in Sardinia at Sarrabus, and there is another at Sainte Marie-aux-Mines in Alsace. The only other localities are at Skulerud, near Drammen in Norway (not far from Kongsberg), and at Tunaberg in Sweden. In the New World the most famous deposit of this type of silver is at Cobalt, Sudbury, Ontario [hence cobalt-type ore]. A small amount is found in the Monte Christo Mine at Wickenberg, Arizona, while the third source is at Batopilas in Mexico.

The Ag-Co-Ni-As deposits occur in igneous rocks, and pure white quartz is invariably present as pegmatite or rock crystal, together with the following important glassforming minerals:

- *Fluorite* CaF_2.
- *Calcite* $CaCO_3$, often stained pink by "cobalt bloom" (erythrite) but also existing in other colors from yellow to blue.
- *Apatite* $Ca_5(Cl,F)(PO_4)_3$, occurring in crystallized concentrations and rich masses in pegmatites and igneous segregations. Average composition is 54.5% Ca, 41.5% P_2O_5, 4% fluorine, and 4% chlorine (note the two important glassformers, phosphorus and fluorine). The violet crystals of Ehrensfriedersdorf in the Erzgebirge are famous, while the brown and green crystals from Sudbury can be up to 18 inches long and three inches wide.
- *Torbernite* $Cu(UO_2)(PO_4)_2.(8-12(H_2O))$, another source of phosphorus, found with the rich silvers of Schneeberg in Saxony. Green crystals of torbernite over one inch across are known.

The granites of these igneous deposits are rich in the isomorphous feldspar series:

- *Orthoclase* $KALSi_3O_8$, with 16.9% K_2O, 18.4% Al_2O_3, 64.7% SiO_2 in which sodium can replace 50% of the potassium in sanidine.
- *Microline* $KALSi_3O_8$, a bright green form of orthoclase. It can grow to crystals several feet in length and is quarried for ceramics and glazes. The plagioclase series, $NaAlSi_3O_8$ to $CaAl_2Si_2O_8$, can contain 11.8% Na_2O in albite.
- *Nepheline* $(Na,K)(Al,Si)_2O_4$, 21.8% Na_2O, occurring in large dull crystals of six inches or more at Bancroft. It fuses easily and has recently become an important glass raw material.

Other interesting silicates form the sodalite group:

- *Sodalite* $Na_4Al_3Si_3O_{12}Cl$, found in rich blue masses with the Ontario silvers and in the Erzgebirge. It contains 25.6% Na_2O, 31.6% Al_2O_3, 37.2% SiO_2, while 7.3% Cl can replace some of the oxygen. It can easily be confused with *Lazurite*, of almost the same chemical composition, $Na_{4-5}Al_3Si_3O_{12}S$ (lapis lazuli).
- *Muscovite* $KAL_2(Si_3Al)O_{10}(OH_1F)_2$, a common rock-forming mineral in pegmatite dykes, the color of which can range from white, yellow, amber, and rose to green. Its composition is 11.8% K_2O, 38.5% Al_2O_3, 45.2% SiO_2, 4.5% H_2O.
- *Biotite* and *Lepidolite*, other abundant rock-forming minerals, rich in potassium (a glassformer) and rich in pegmatites (a source of silica).

All these minerals are found in quantity with the Ag-Co-Ni-As group. They are, in fact, Agricola's "solidified juices" which will melt again with heat, and his "stones which easily melt in the fire" (Hoover 1912:380) (see pl. 2).* He describes these stones in detail in *Rerum Metallicarum Interpretatio* (1546; Hoover 1912:380), in *Bermannus* (ibid., 381), and in *De Natura Fossilium* (ibid., 380). Agricola states that when thrown into hot furnaces, these stones *flow:*

The first type [of stones] resemble transparent gems but are resplendent. Some resemble crystal, others emerald, heliotrope, lapis lazuli, amethyst, sapphire, ruby, and other gems, but they differ from them in hardness.

Agricola is obviously describing the fluorite family here (see pl. 2d) which are glassformers.

The second type is not translucent, and abounds in veins of its own. The second order of stones does not show a great variety of colours, and seldom beautiful ones, for it is generally white, whiteish, greyish or yellowish. These stones melt very readily in the fire and are added to ores from which metals are smelted.

Here Agricola is describing the apatites, nepheline, torbernite, microline, sodalite, and muscovite indicated above, and also erythrite, the pink ore of weathered cobalt, common in Saxony. Four-inch crystals of this raspberry red mineral in quartz are common at Schneeberg, while borate crystals (another glassformer) are also found in these high temperature veins (see pl. 2a,b,c).

The third order is the material from which glass is made, although it can be made from the other two:

The third order is white quartz pebbles from the rivers where tin is collected—the whiter the better [pl. 1c]. . . . None of these stones contain any metal.

In *Bermannus* (Agricola 1530; Hoover 1912:381) there is a revealing conversation.

Bermannus: "You see the other kind, a paler purple colour?"

Naevius: "They appear to be an inferior kind of amethyst, such as are found in many places in Bohemia."

*Agricola was a famous sixteenth-century historian, physician, and pioneer in mineralogy.

Bermannus: "Indeed they are not very dissimilar; therefore, the common people who do not know amethyst well, set them in rings for gems, and they are easily sold . . . It is indeed not far from here [Joachimsthal at Breitenbrunn which is near Schwartzenberg] . . . Moreover from *fluores* [slag] they can make colours which artists use."

These would be purple fluorite crystals found in the silver veins. Pierre Weidenhammer of Schneeberg is said to have been the discoverer of *zaffre* or cobalt blue in 1520. Also in that year, rich, native silver-ore bodies were found at Schneeberg, just below the surface, and worked without the need for lamps. Curiously, Agricola, who lived so near at Joachimsthal, does not mention these facts. However, on the evidence of Leonardo da Vinci's painting of 1504, "The Virgin on the Rocks," the pigment must have been discovered before 1520. In fact, cobalt blue glass was used in cathedral windows as early as the eleventh and twelfth centuries at Augsberg, Le Mans, and at Canterbury where it is dated to 1180 (see color pl. I opposite p. 292 in the *Encyclopaedia Britannica*, 1963 ed., vol. 21).

Biringuccio, in *De la Pirotechnica* (1540, Book II, ch. 1:x), was the first author to describe *zaffera* (from sapphire-sappheiros). He mentioned it as

another mineral substance [undoubtedly smaltite, smalt-blue frit], like a metal of middle weight, which will rot alone, but accompanied by vitreous substances it melts into an azure colour so that those who colour glass, or paint vases or glazed earthenware, make use of it. . . . If one uses too great a quantity of it, it will be black.

Here is a perfect description of the coloring power of cobalt (see pl. 2e). Antonio Neri in *L'Arte Vetraria* (1612) knew that blue glass came from cobalt, and Christopher Merrett in *The Art of Glass* (1662) stated that cobalt blue glass came from Germany. In 1679, Johann Kunckel specified Schneeberg as a center of cobalt glass production (*Ars Vitraria Experimentalis*). Agricola, who certainly used Biringuccio's work and may well have met him when he was in Italy in 1524–1526, does not use the word *zaffre*, and completely missed the connection between it and cobalt. He is confused between cadmia, spodos, and bismuth, and says that "the slag of bismuth mixed together with metalliferous substances, which when melted make a kind of glass, will tint glass and earthenware vessels blue." Bismuth here must have meant cobaltite or smaltite, a significant name. Again, curiously, Agricola never mentioned the striking pink crystals of erythrite, for which Schneeberg was famous.

Hoover, in his translation and commentary on *De Re Metallica* (1912), also completely missed the geological significance of the associated stones, especially fluorite and apatite, together with those outlined above, in silver smelting, and is responsible for two very misleading statements.* The first is "that early ancient copper must have been smelted" (1912:353n) as distinct from melted. Enormous quantities of native copper have been found in northern Michigan (Pearl 1966:178) and Cornwall (see Dayton 1978, fig. 11). Native copper was extremely abundant to ancient man and also remarkably pure and suitable for melting and elaborate theories of the *early* smelting of copper carbonates and sulfides must be viewed with suspicion. Next, after native copper, much cuprite or ruby copper (88.8% Cu, 11.2% O) was found associated with it near the surface, and this copper ore is very easy to reduce to copper with charcoal, and would have been one of the ores first smelted by early man.

There are two main types of metal ores, as distinct from *native* metal ores, i.e., pure copper, silver, gold, etc. The *oxides* are found near the surface, above the water table, and do not require roasting. They are easily smelted into metal with charcoal, a flux, and a silica to form a slag. The *sulfide* ores are rich in sulfur and these have to be roasted first to drive off the sulfur, otherwise they cannot be smelted. The preliminary roasting is difficult, especially since copper has a great affinity for sulfur. The sulfides were usually left alone by early man.

The second erroneous statement that Hoover makes is that "silver does not appear native in any such quantity, [so] we must assume a knowledge of cupellation for the parting of the two metals" [i.e.,

gold and silver] (Hoover 1912:390). Silver was found native in very large amounts until 1914 (see pl. 1b). In 1890 a nugget weighing 1,840 pounds and containing 90% silver was located in the Mollie Gibson Mine at Aspen, Colorado. The La Rose Mine at Cobalt, Ontario, had a "silver sidewalk" of almost solid silver, 100 feet long and 60 feet thick. A slab of it weighing 1,640 pounds can be seen in the Parliament Building in Toronto. This silver was from the classic Ag-Co-Ni-As-type deposit in Ontario. In the similar type of deposit in the Erzgebirge, Agricola (Hoover 1912:76) details the enormous quantities dug out there up to his time:

> Albertham [the Gottesgab mine] . . . for they have dug out of it a large quantity of pure silver, 300,000 ounces, and a vein at Annaberg, called Himmelsch Hoz, which surpasses all others, where some 420,000 ounces were produced . . . But far above all others is the Saint George Mine at Schneeberg with two million ounces.

It was in the Saint George Mine, in 1477, that a mass of silver and argentite was found weighing 20 tons. The mines of Schneeberg had reopened in 1471 and a color works was set up (which continued to 1914). This was to exploit the natural blue slag accidentally produced from the smelting of the rare Schneeberg silver. But the Schneeberg deposit must have been worked, on and off, since 1800 B.C. on the evidence of the blue cobalt glass found in the ancient world.

Joachimsthal had been founded in A.D. 1516, and from 1775 to 1795 the ores there averaged 78% silver. Agricola describes rich pockets or "glory holes," often measuring 600 feet by 600 feet.

The mines of Freiberg, 50 miles away, had been worked as early as A.D. 1200, and were probably known by the Romans. Agricola provides a hint that these mines were worked earlier, for he describes (Hoover 1912:42) a vein at Annaberg which was called the Kölergang because a charcoal burner had been discovered in it. Altenberg in Saxony (which the writer visited in 1985) has one of the largest tin mines in the world where tin outcropped on the surface. These surface workings collapsed, leaving a huge crater or "pinge" exposed on the side of the hill, but tin is still being mined there today. It is most unlikely that such a large surface outcrop would have escaped the attention of early man (Mantell 1949).

As already stated, the Ag-Co-Ni-As group can occur in veins five feet wide and does not contain

---

*It appears from the writings of Biringuccio and Agricola that a large-scale development in ore-smelting took place during the Renaissance, perhaps inspired by the writings of the Greeks, Romans, Moors, and Arabs, and by the invention of printing in 1440. Agricola is confused in many sections, and Hoover (1912) is rather weak on geology and chemistry, which is not to decry his great translation and commentary.

MAP 3

Ore Deposits and Trade Routes

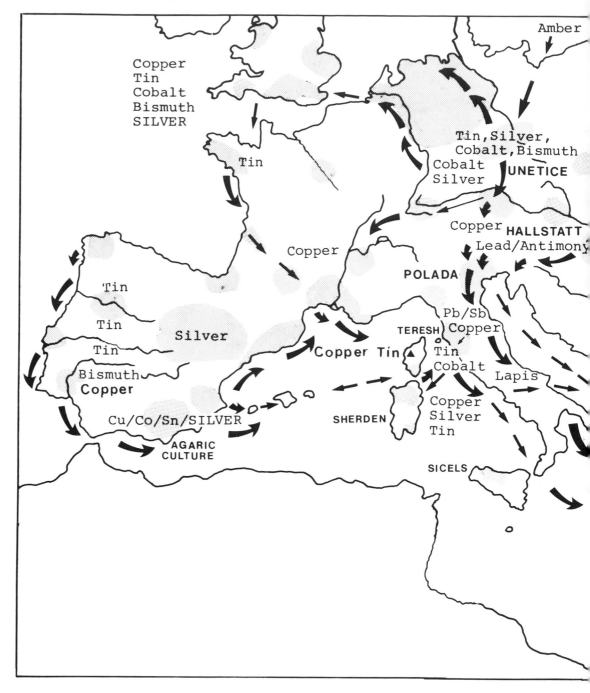

This map shows the mineral deposits of Europe and the areas inhabited by the Beaker People (shaded). Cobalt deposits were even more scarce than the rare tin. The chief cobalt deposits of Europe were in the western Erzgebirge at Schneeberg. These mines provided Europe with cobalt blue glass for 800 years until 1914. Elsewhere, cobalt was found at Anarak, Zaire, and Queensland (without silver), but at Cobalt, Ontario, it was associated with massive native silver.

From this map it appears that the Danube was of minor importance in the transportation of metals to the Near East, while the sea routes of the Adriatic and from Spain via the Balearics, Corsica, and Sardinia were of great importance.

MAP 3

Ore Deposits and Trade Routes *continued*

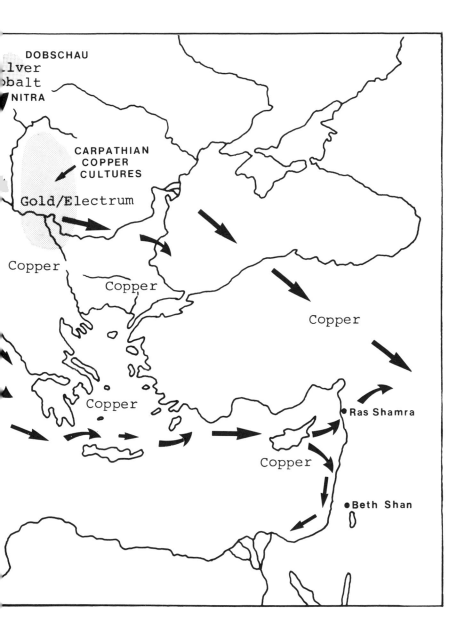

DOBSCHAU
lver
balt
NITRA

CARPATHIAN
COPPER
CULTURES

Gold/Electrum

Copper

Copper

Copper

Copper

Ras Shamra

Copper

Beth Shan

The distribution of the Beaker cultures over western Europe. This culture marked the beginning of metallurgy, and Carbon 14 dates indicate around 1900 B.C. for its floruit. The sites were noticeably located in metal-rich European areas of copper and tin. The "Maritime Beaker" group extended along the Atlantic seaboard of Portugal and Spain, and across the Western Mediterranean via the Balearics and Sardinia to Tuscany. The related Únêtice culture, with its similar quantity of tin bronze, was located near the tin and copper deposits of the Erzgebirge, from which the culture spread southwards into northern Italy, after a climatic warming ca. 1800 B.C. which opened the Brenner Pass.

In the Proto-Bronze Age, the ubiquitous Beaker Folk were present throughout western Europe.

galena or gold. What is more, the weathered surface ores are extremely rich, containing native silver (100% Ag), argentite Ag_3S (87% Ag = 25,000 ounces per U.S. ton), and cerargyrite—"horn silver"—AgCl (75% Ag). Often found in seven-pound lumps, this ore has produced more than two billion ounces of silver at Potosi, Bolivia, since 1544 (Beyschlag 1914:644–649, 680), while argentite provided most of the 400 million ounces of silver from the Comstock Lode in Nevada. Cerargyrite forms with bromine as bromyrite AgBr, which is very rich and easily smelted.

In 1841 at Chañarcillo a mass of bromyrite and cerargyrite was found weighing 45,000 pounds, which produced 30,000 pounds of pure silver. Next in richness come the "ruby silvers," proustite Ag_3AsS_3 (65% Ag) and pyrargyrite Ag_3SbS_3 (60% Ag), grading 19,000 to 17,000 ounces per ton. Then comes stephanite, AgSb, at 68% silver. These ores could be described as recently as 1907 "as the most important silver ores" (Brown 1907:182). Brown goes on to state that "galena is not rich in silver, 60 ounces per ton, sometimes even 300 ounces, and rarely 1,500 ounces." It confirms this writer's analyses of many hundreds of galenas from the Old World which contained no silver *at all* and which contradict the common statement that "all galenas contain some silver." Obviously, compared with the rich, dry ores (i.e., those without lead), which were easily smelted, a silver-rich galena of 300 ounces per ton was not worth troubling with, considering the enormous amount of wood needed to drive off the sulfur first, and then the lead by cupellation. Even as late as 1968, galena only provided some 40% of the world's silver (Boegel 1971).

Cupellation is an assaying technique, done with very small amounts of an ore sample to see how rich the ore is, and if it is worth further processing. Because early archaeological objects of silver were very pure, archaeologists thought wrongly that the ancients were using the very sophisticated cupellation technique to produce their silver. Whereas, in fact, they were using native silver (or copper) or the rich, easily smelted oxide ores which are found on the surface.

Native silver *not* of the Ag-Co type was found on the surface at Tonopah, Nevada, where "bonanza" ore ran at 121.08 ounces silver with 1.50 ounces of gold per ton (see pl. 1a). At Tombstone, Arizona, one can see the sky a few feet overhead from the famous rich glory hole, which began about three feet below the surface. At Butte, Montana, which now has the largest hole in the world, the surface iron-hat, rich in gold, consisted of manganese silicates; the next 150 feet of oxide zone was rich in silver and gold; then, in the cementation zone at 220 feet deep the silvers ran at 100 ounces a ton, containing gold at 0.2 ounces. Below this, in the huge sulfide mass, the ore ran at 30 ounces silver, 0.03 ounces gold, 2.3% lead, 1.0% zinc, and 0.6% copper. On the surface, the northeast veins extended for eight miles! After the boom days the mine became principally a copper producer. This situation is typical of many of the native silver mines of the world, with the exception of the Ag-Co-Ni-As group.

The mines of Laurion in Greece are first mentioned in the Athenian records of 500 B.C., when the famous owl coins were produced, and worked until 413 B.C. They produced about 84,000 ounces of silver in 484 B.C.. The earlier *Wappenmünzen* coins of Athens have been shown, by the writer's lead isotope analyses (Dayton 1978:441), not to have come from Laurion ore, while the turtle coins of nearby Aegina were of Spanish silver. In 300 B.C. the Laurion mines were opened again by the Athenians and the supporting pillars of ore were robbed to aid in their war with Sparta. Three hundred years later, Strabo wrote (IX:1, 23), "The silver mines of Attica . . . are now exhausted. The workmen, when the mines yielded a bad return, committed to the furnace the old refuse and scoria, and thence obtained very pure silver, for the former workmen had carried on the process in the furnace unskillfully."

Ardaillon (1897) states significantly, that when the old slags and *rejected ore*, estimated at seven million tons, were reworked by a French company after A.D. 1860, 40 to 90 ounces of silver per ton were recovered. What he does not tell us is how much was slag and how much was rejected ore. *Why* was the ore rejected, and how did the Athenians know its grade of silver? The writer would suggest, no evidence remaining, that the rejected ore was the crushed and washed galena from which the free silver had been separated. The Athenians certainly knew that the upper vein of galena, at 80 ounces per ton, was not worth working and they did not mine it. An experienced miner can tell a silver-rich galena by eye, because it has a bluish color.* At Laurion the lower contact ran

---

*In confirmation of this theory, this author recently obtained a rare old sample of argentiferous galena from Wheal Herland, Gwinear, Cornwall, where native silver can clearly be seen associated with the cubic crystals of galena.

at 180 ounces per ton, and it was the waste from this ore that yielded the French company its 90 ounces per ton of silver.

Cupellation was probably not practiced in the early days of Laurion. Since silver-rich galena usually contains *free* native silver and argentite at the crystal planes, giving the galena a purple-blue luster, this silver was liberated by the crushing and washing processes, the remains of which can be seen at nearby Thorikos. Thus, the yield would be a concentrate between 100% and 87% silver. The specific gravity of silver is 10.49 at 15°C, while that of galena is 7.5, showing that washing could separate the silver and argentite from the galena.

The rich yield in 484 B.C. could also have come from native silver wires in the cerussite ($PbCO_3$) in the weathered oxide zone. Silver melts at 960.5°C, and the concentrates mentioned above could have been smelted in the simple furnace found by this author at Laurion (Dayton 1978:111). Silver may well have been the first ore to have been smelted, as copper is much more difficult to reduce because of its affinity for sulfur and its higher melting point of 1083°C.

Charles Sewell, who carried out a geological survey of the old Kassandra Mine in Chalcidice, northern Greece, reports that his team found ample evidence of ancient workings, all of which stopped at the sulfide zone (personal communication). This mine could well have been the legendary one worked by Pisistratus, Tyrant of Athens (died 527 B.C.), and later by Philip II and Alexander the Great. The sulfide ore from Kassandra examined by this author was rich in gold, copper, and bismuth, and very rich in silver and lead.

It is certain that the Athenians did not know of the Parkes Process of the early nineteenth century, when zinc was deliberately added to attract the silver, in preference to the lead. The silver-rich concentrate then was roasted to drive off the zinc and leave the pure silver (see Hofman 1918). The free silver in the galena at Laurion was probably separated by gravity and then roasted, since the most remarkable remains of mining at Laurion are the extraordinary complexes of washeries, with channels and settling tanks, which have been excavated in recent years and can be visited. Unlike the Romans, the Ancient Greeks do not appear to have had any use for lead.

MAP 4

The Unity of the European Bronze Age

KEY.

✪ Major metal deposits worked in ancient times in Central Europe.

▲ Mycenaean finds

■ Bronze Age sites & hoards

● Phoenician sites

The evolution of the Bronze Age, ca. 1800 B.C., in three areas is clear:

1. The Erzgebirge for copper and tin bronzes; 2. Pribram, with copper/antimony bronzes; 3. Dobschau for copper with cobalt traces and blue beads.

From these areas three easy routes run via the Brenner, Predil, and Postojna passes to the rich sites of the Po Valley and on to the key site of Bologna—the crossroads of Italy.

The Balkan Copper cultures form a separate group. The Danube is not as important a trade route as the Adriatic; Sardinia has links to Iberia; and Tuscany forms a third distinct culture.

It is clear that the "so-called" Mycenaean/Minoan group is very much a coastal and sea-faring culture, trading via the Adriatic with northern Italy for Central European glass and metals and Baltic amber; and *also* with Sardinia and Tuscany. Note how the Mycenaeans do not extend into northern Greece and that the sea crossings from Italy to Albania and from the Peloponnese to Crete are each a mere 50 miles. Northward from Únêtice a thick chain of sites leads down the Elbe to the amber of the Baltic and to Holland and Britain.

Finally, after a curious gap of 800 years, the Phoenicians are trading over the same routes for the metals and glass of the Hallstatt and La Tène cultures.

1. WESTERN ERZGEBIRGE: native silver, cobalt, and bismuth.
2. CENTRAL ERZGEBIRGE: mixed tin and copper, red jasper.
3. EASTERN ERZGEBIRGE: tin, silver, lead, copper, zinc.
   (The whole Erzgebirge is rich in amethyst, rock crystal, jasper, fluorite, apatite—all materials for glassmaking.)
4. PRIBRAM: silver, lead, zinc, copper, and antimony.
5. KREMNICA: native silver with a little gold.
6. DOBSCHAU: copper with strong cobalt.
7. BISCHOFSHOFEN: copper with traces of silver and antimony.
8. BLEIBERG: lead/zinc, BLUE ANHYDRITE for 12th Dynasty Egypt.
9. KAPNIK: classical antimony ores with copper and gold.
10. APUSINI MOUNTAINS (DACIA): Roman gold and silver mines.
11. THE BANAT.
12. RUDNIK.
13. BOR: copper/lead deposits.
14. TREPCA: lead/zinc.
15. TUSCANY: native antimony.
16. TUSCANY: Native copper, tin traces.
17. SARDINIA: Calabona, copper.
18. FUNTANA RAMINOSA: copper, lead-zinc, traces silver, bismuth, cobalt.
19. IGLESIAS: silver, tin, lead/zinc, bismuth, traces cobalt.
20. CASSANDRA AREA: chalcopyrite, lead/zinc, gold.
21. LAURION: much copper, argentiferous lead, zinc.

3 EXPERIMENTS AND ANCIENT METHODS OF METALLURGY

Having deduced where and how cobalt glass was made, the problem remained to prove the thesis empirically. The rich Ag-Co-Ni-As ores are now museum specimens, and if obtainable, worth $200.00 an ounce or more. Samples were acquired over the years: some through the kindness of the British Museum [Natural History], others from dealers and rock-hounds, and some on field trips by the writer to places such as Mexico, Arizona, and Sardinia, etc. Through a valuable contact samples were collected from Schneeberg, Ehrensfrieders-dorf, and other mining towns in eastern Germany and Czechoslovakia. The specimens were too valuable and too small to use for smelting experiments.

Types of Ores

More than 90 samples were analyzed and all proved to be very rich in silver, nickel, cobalt, and bismuth. One silvery-looking example from Schneeberg indeed proved to be native bismuth with strong traces of nickel and cobalt, but with no traces of silver or lead. Lead was found only in one sample from Jachymov, Czechoslovakia, at a level of 50 parts per million, but traces of copper occurred in some samples, explaining the weak copper in some cobalt glasses and faience. On a field trip to Batopilas in Mexico in 1981, similar samples were obtained where cobalt was found to be in the gangue of pink calcite.

Methods of Smelting

Finally, in Creede, Colorado, in 1983, a ten-ounce sample of rich, dry ore was obtained, and an assay ton (29.166 grams) was assayed by cupellation at 3,810 ounces per ton. The ore was mainly native silver in wire form on a quartz gangue, and the slag produced by the crucible process before cupellation was a dull and uninteresting black color because of the added litharge, which shows that a lead ore would never have produced a cobalt blue glass. The assay was carried out in Nevada by a very experienced assayer and was the richest he had ever encountered. Another sample of the ore was mixed with 1 part sodium bicarbonate, 5 parts litharge, 1 part borax, ½ part fluorite, and a teaspoonful of wheat flour, to produce a fluid slag. (Silica flour is added if the ore does not contain enough silica to form a glassy slag, which carries off the impurities.) The fused mixture was poured into a conical iron mold. When cooled, the lead button was obtained for the next stage, cupellation (pl. 5b,c,d). Temperature control in the furnace for cupellation is critical at 926°C (see fig. 2b). If much higher, the silver will begin to volatilize, and if below 807°C, the button will freeze and the process cannot be started again.

To another ounce of the finely ground Creede ore, 0.25% of cobalt oxide was added in order to produce an artificial mixture corresponding to the Schneeberg and Batopilas ores. Fifty grams of sodium bicarbonate, 30 grams of silica, 3.5 grams of borax, a teaspoonful of wheat flour, and 5 grams of fluorspar—all finely ground—were added to the mixture. The charge was gently fused in a Colorado-type crucible (pl. 4b). A salt cover was not thought necessary, and no lead or litharge was added. After 40 minutes the mixture was fused (i.e., liquid) and poured into a preheated conical iron mold to cool (pl. 5b). Some of the slag, which was a dark, navy blue cobalt color, was poured onto an iron plate. After five minutes of cooling, the drops suddenly exploded with a great noise into many tiny fragments. The slag of the second charge, in which the sodium bicarbonate had been increased to 65 grams, was poured onto a preheated iron plate and cooled successfully without shattering.

The third charge contained 40 grams of sodium bicarbonate and no borax. In the dark blue color of the resulting slags, all three were identical, and all three produced a button of silver. It must be stressed that an experienced assayer can tell at a glance what fluxes are needed for different types of ore. Note that no litharge (PbO) was used in the above two experiments, nor in the following two, although a button of silver was obtained in each case.

Two more charges were then prepared with 0.125 grams of cobalt oxide, each with 30 grams of silica, 5 grams of fluorspar, and 65 and 40 grams respectively of sodium bicarbonate. Both produced silver buttons and a beautifully rich cobalt blue slag after fusion in a crucible after some 30 minutes (pl. 4b,c,d). The whole process was remarkably easy, and the following were observed:

1. A very low percentage of cobalt oxide was required to produce the blue color, showing that only traces of cobalt in the gangue or the ore were needed.
2. Silica in the form of quartz *had* to be present in the gangue if a glass slag was to be produced (i.e., a limestone gangue, galena, or silver ores in limestone would not produce a glassy slag).
3. Cupellation, a more complex process, would not have been needed to produce a silver button from such rich, dry ore. This process for dealing with silver-rich galena is explained later.
4. In the second stage, when the slag was remelted to make glass beads, tongs would have been vital to hold the red-hot crucible and to pour the glass into bead molds.

From these observations it was deduced that early man had to have known the shock-resisting qualities of steatite molds, found at Mycenae, as well as the necessity of preheating them or using an annealing oven. The required clay crucibles would have been available to early man: they would have been smashed when cold, at the end of the fusion process, and the silver button separated from the blue glass slag (see pl. 5c). Perhaps he also knew of the deliberate use of fluxes, which occurred naturally in the gangue at Schneeberg: in this case, fluorite, apatite, and feldspar (see pl. 2a–d). He may have used common salt, abundant in the Hallstatt region, to serve as a cover to form the glassy slag and as a wash to bring down adhering material from the side of the crucible. Charcoal might also have been used in the charge and potash obtained from ashes. Early smelting sites should be littered with masses of fragments of glazed crucibles, used once and broken to separate the metal from the slag. If cupellation *was* practiced, thousands of used cupels should be strewn about, for a cupel can be used only once (pl. 5e). All these artifacts are found in small mountains around mining camps, together with equal heaps of black glass slag

(see pl. 3), and archaeologists should be on the lookout for them.

The writer has found no cupels or crucible fragments on several field trips to Laurion, but cobalt blue glass slag has, in fact, been found at Schneeberg (pl. 3e). Agricola casually mentioned (Hoover 1912:400) that native silver and "rudis" silver (the rich silver ores described above) are not smelted in a blast furnace, but in small iron pans or in a covered pot in a furnace. There was no need for any of the Erzgebirge rich silvers to be smelted with lead, which was added as the oxide, PbO (litharge), to low grade ores to collect the silver.

It appears that the use of bone ash was relatively new in Agricola's time, for he stated (Hoover 1912:247): "Once upon a time the base metals were burned up, in order that the precious metals should be left pure; the Ancients even discovered by such burning what portion of gold was contained in the silver, and in this way *all the silver* was consumed, which was no small loss" [author's italics]. This is reminiscent of how the Kassite King Burnaburiash II, in about 1370 B.C., tested the gold sent to him by the Pharoah Akhenaten "in the fire and found it wanting."* He probably burned off the copper and silver in the gold.

The Romans, however, certainly knew how to desilver lead, for lead ingots stamped "ex argent" are common. They may have done this by adding a zinc ore which forms a dross or crust on the surface. This process collects the gold, silver, and copper in the molten lead; then the crust is skimmed off and treated. Or, they may have used what was later known as Pattinson's Process.

Here, the lead containing silver is melted in a series of pots, perhaps as many as 20. When the lead in a silver-rich galena, containing below 896 ounces of silver per ton, slowly cools below 327°C, small crystals start to form on the surface. These are *richer in lead* than the remaining molten part. The crystals then sink to the bottom of the pot, where soon they are found in quantity. The lead-rich crystals are then removed by straining the molten lead with a perforated iron scoop and transferred to the next pot in the row. In this way the pots at one end will eventually contain pure lead with a little silver in it, while the pot at the other end will contain silver-rich lead, in practice not much more than 600 ounces per ton. This silver-rich lead is then transferred to another dish-like crucible, and

---

*From a clay tablet found at Tel el Amarna in Egypt.

FIGURE 2

The Cupellation Myth

A—Muffle. B—Its little windows. C—Its little bridge. D—Bricks. E—Iron
door. F—Its little window. G—Bellows. H—Hammer-chisel. I—Iron ring
which some use instead of the test. K—Pestle with which the ashes placed in
the ring are pounded.

a

b

2a. The illustration above from Agricola shows a typical advanced cupellation furnace around A.D. 1520 with muffles and the vital tongs. Cupellation was a precise analytical technique to determine the grade of ore, and not for the production of silver per se. Why, therefore, have not masses of slag, muffles, cupels, and even furnaces been found at all assumed smelting sites, as at Laurion and Rio Tinto ? It will be shown that the process illustrated is highly sophisticated.

2b. Here, the assayer is carefully looking at the molten lead in the cupel in the muffle through a slit in the wooden board to see the "blick" stage when the last of the lead boils off, leaving the silver bead. The cupel must be moved at this stage and the process stopped or the silver will boil off in its turn and be lost.

FIGURE 3

The Cupellation Myth *continued*

A—Rectangular stones. B—Sole-stone. C—Air-holes. D—Internal walls.
E—Dome. F—Crucible. G—Bands. H—Bars. I—Apertures in the dome.
K—Lid of the dome. L—Rings. M—Pipes. N—Valves. O—Chains.

Agricola

3. A large-scale silver smelting furnace with a solid brick base and iron cover. If silver was being produced on this scale in ancient times, surely the remains of such furnaces would exist. The base of the furnace (F) would have been made from bone ash mixed with beer or milk as a binder to absorb any lead present, as distinct from deliberately added litharge if mixed ore of "dry" silver and argentiferous galena was being smelted, as at Joachimsthal.

FIGURE 4

The Cupellation Myth *continued*

A—FURNACE. B—STICKS OF WOOD. C—LITHARGE. D—PLATE. E—THE FOREMAN
WHEN HUNGRY EATS BUTTER, THAT THE POISON WHICH THE CRUCIBLE EXHALES MAY NOT
HARM HIM, FOR THIS IS A SPECIAL REMEDY AGAINST THAT POISON.

Agricola

4. The large-scale furnace in action. The fumes were probably of arsenic, and the
smelter's life would have been fairly short. He is tapping the molten slag (not
litharge) from the furnace. The end product would have been a cake of crude blister
silver which would have been further fire-refined in the closed pot that the woman
is carefully finishing.

FIGURE 5

Ancient Smelting Fluxes

a

After Erker 1574

b

5a. Fluxes are vital for successful smelting and serve to make the slag liquid and flow, and so separate the impurities from the metal to be produced from the ore. Potash, soda, fluorite, and borax are excellent fluxes. Agricola mentions "stones which melt in the fire," meaning fluorite and apatite. In the Middle Ages the production of saltpeter was a vital industry. Saltpeter was produced from clamps of nightsoil and animal manure, which were left for some five years, as seen here. The clamps were then washed and the liquor boiled down to a salt, rich in sodium, potassium, and phosphorus.

5b. Here, the workers are carefully scraping efflorescence off the walls to use as saltpeter. It is not known how early man discovered fluxes and what they were, but their discovery represented a significant breakthrough in metallurgy. Is it possible that native silver associated with fluorite crystals (as in pl. 2) led to the discovery of fluxes, as in the case of silver smelting?

heated to below the boiling point of silver (960.5°C). The lead is then oxidized away with powerful blasts of air from a bellows into a liquid slag [such as litharge], leaving a bun of silver, or silver and gold. Bismuth also goes into the silver and gold.

Cupellation

Agricola gave excellent treatment to the subject of cupellation (see pl. 5) and to the extraction of silver from copper and from galena.* He obviously knew well the "Sternen" silver/lead mine at Joachimsthal, since he was the local doctor there before he moved to Meissen. The Sternen had produced, according to him, 350,000 ounces of silver. It must have been from this mine that he derived his knowledge of cupellation (not needed with the 20,000 ounces of dry silver ores of Schneeberg; see fig. 2a,b) and the large-scale roasting in what was really a large cupel with a three- to five-ton charge (illustrated in Hoover 1912:468, 470, 474; also see figs. 3, 4). In fact, most of his work dealt with the desulfurizing of galena and copper. It is of interest that Agricola possessed a copy of Theophilus' *De Diversis Artibus* (Dodwell 1961) and may well have perpetuated alchemic and mystical processes.

Agricola accurately described the cupellation process. When dealing with silver-containing galena, it is essential to know the exact amount of silver in the ore, in order to avoid wasting labor and fuel. As it is a delicate *assaying* technique (not a smelting process), accurate scales are needed. Today, the method has barely changed since Agricola's time 450 years ago. A constant amount of crushed ore is carefully weighed [29.166 grams for a short ton and 32.666 grams for a British ton]. Depending on the gangue minerals with the ore [i.e., if it is acid with much silica, or basic with little silica and much iron oxide and limestone], extra silica is added in the form of crushed quartz or rock crystal. This process forms a good glassy slag which, in floating like fat on a soup kettle, removes the impurities from the ore. Fluxes are added to assist the formation of the glass slag: sodium, potassium, borax, apatite (which contains phosphorus), and fluorite in their various forms. The soda ash used today (Na_2CO_3) melts at 852°C

and lowers the melting point of silica from 1750°C to below 1000°C and even lower. Borax glass does the same, melting at 742°C. An excess of borax will form a matte and attack the crucible. As explained above, fluxes are normally abundant in the gangue of the Ag-Co-Ni-As silver deposits and so would not have had to be deliberately added at Schneeberg to the rich, iron-free silver ores, much of which were silver chloride (AgCl). These fluxes are Agricola's "stones which melt in the fire" (Hoover 1912:380). If lead or galena is present [and it is not found in the Ag-Co group], a reducing agent must be used in a manner which, from nineteenth-century books on assaying, reads like a cookery recipe.* The amount of reducing agent must be calculated according to its power, e.g.:

- 1 part of wood charcoal will reduce 22 to 30 parts of metallic lead
- 1 part of wheat flour will reduce 15 parts of metallic lead
- 1 part pulverized white sugar will reduce 14.5 parts of metallic lead
- 1 part argol (crude bitartrate of potash) will reduce 5.5 to 8.5 parts of lead
- 1 part cream of tartar (purified argol) will reduce 4.5 to 6.5 parts lead

The Greeks and Romans (Dioscorides:XXIII; Pliny:31) knew argol as a deposit in wine casks or vats during the fermentation process. It could also be obtained from the crushed grape waste used to make white lead oxide. Duhamel did not identify potassium and sodium until the eighteenth century, so one should not think that ancient "soda" is always sodium carbonate.

Nor must one forget the organic materials used by early man, animal blood and the ashes made from the scraping of hides, both rich in potash. In the Middle Ages large saltpeter plantations were common outside every town (fig. 5). Night soil and animal manure were placed in long lines, covered with earth and left for five or six years. The earth was then washed and the filtrate, if it tasted salty, boiled down to crystals of argol and used for gun-

---

*He did not appear to have seen or studied the methods which were used for treating the rich, dry silver ores some 60 kilometers north of Joachimsthal on the other side of the Erzgebirge, or he would surely have described the color works of Schneeberg.

---

*In a recent visit to a silver smelter in Australia, the same sort of recipe was chalked up on a blackboard as in Agricola's time:

2 bags borax
1 bag soda
2 scoops MnO_2
2 scoops nitrate

powder. *Sal Artifiosus*, similar to argol and known to the Romans, was made in various ways:

1. Equal parts of argol, lees of vinegar [more potash] and urine [rich in potash, phosphorus, sodium, and, when stale, ammonia] were all boiled down to a salt.
2. A pound each of the ashes which wool-dyers use [the grease of sheeps' wool is rich in potassium], argol, lime, and melted salt was thrown into 20 pounds of urine. The mixture was boiled down to one third and strained; then 1 pound of salt was added, along with 8 pounds of lye (potash, K_2CO_3, from wood ash), and the whole was finally boiled down to salt.
3. Salt and rusty iron were put in a vessel which was filled with urine, and left in a warm place for 30 days. Phosphorus and ammonia would form in this time: phosphorus was first discovered and produced in A.D. 1667 from stale urine, a method which continued to be used for the next 150 years.*

The above ingredients were mixed with the ore in *a crucible* for the first part of the cupellation process, together with a carefully calculated amount of the much misunderstood material litharge, PbO, varying from 1.5 to 5 grams if the ore was of the basic (silica-poor), lead-rich Laurion type. The purpose of the litharge was to draw down the silver and gold in the ore into a button of lead at the bottom of the slag. Bismuth also passed into the button, while antimony and arsenic volatilized, and cobalt passed into and colored the slag. Sometimes the base metals, antimony, iron, and copper, united with the excess sulfur, if the quantity of the reducing agent was not sufficient, and formed a matte between the button of lead and the slag (fig. 9b).

Matte, an artificial sulfide of one or more metals, forms when the charge is too acid (i.e., contains too much silica). Matte always carries some gold and silver. Extra charcoal is needed or an iron nail inserted in the charge. The quantity of charcoal used with any particular ore is determined by *guessing*

its amount of sulfur, in order to get the right lead button. If there is much sulfur, as in galena, zinc blende, and iron pyrites, the ore may require scorification or pre-roasting. The colors in the interior of the clay scorifying dish indicate the minerals in the ore. A green becoming brown at the edge indicates copper; a black through rich mahogany to a light yellow-brown, with a red-brown always present, indicates iron; shades of lemon yellow, lead; purple-black to light violet-brown and amethyst, manganese; and a beautiful blue, cobalt. Early man probably discovered colored glass this way. Cobalt in a *cupel* (as opposed to a scorifying dish) is brown, and therefore early cobalt blue glass could not have been produced in a cupel, confirming that early silver-with-cobalt ore in quartz was scorified in an open clay dish. Again, glazed sherds of these dishes should exist in great quantities at ancient smelting sites and archaeologists should be on the lookout for them. (See pl. 3 for different colored smelting slags collected by the author.)

Note that some ores are more or less reducing, bringing down too much lead, while others are oxidizing, not bringing down any lead and also preventing the reducing agents from acting. Ores with an oxidizing action are red oxide of iron, red oxide of lead, black oxide of copper, and black oxide of manganese. These are all found near the surface at Laurion and would have helped to remove the lead from the silver. Arsenates and antimonates are liable to keep silver away from the lead button; liquid metal arsenide forms a speiss which dissolves any gold present [charcoal helps with arsenical ores]; while zinc ores present (as at Laurion) may volatilize the silver *or* retain it in the slag.

Litharge, a flux, is the monoxide and yellow oxide (PbO) of lead. It melts at 883°C, is unstable, and if exposed to air, rapidly turns to red lead (Pb_3O_4). This causes oxidation of silver and, therefore, loss. For this reason, litharge must be kept in an air-tight container [and one asks how early man did this?]. Cerussite (PbCO), however, which often contains wires of native silver and is found in the weathered surface of galena deposits, is quite a good substitute for litharge, and was probably used accidentally at Laurion.

It is important to note that the metalliferous minerals in an ore may either reduce (take away oxygen from the litharge) or oxidize (give oxygen to the reducing agent). Sulfur, arsenic, antimony, and zinc present problems, as they are reducing agents. Oxides of iron, lead, copper, and manganese, in their highest forms of oxidation (i.e., surface ores

---

*Urine was certainly not wasted in the Middle Ages, when we read that "the pisse of two olde men suffices to fertilize an acre of garden" (Baron von Liebig), while its collection in Ancient Rome was a valuable state monopoly. It was an important bleach, and even today is used by Bedouin women to bleach their hair and to make dyes with rock alum and wild mushrooms.

as at Laurion) give up some of their oxygen in the crucible, are converted into lower oxides, and go into the slag. The sulfides (apart from AgS) also create great difficulties and could not have been used by early man, as shown at Kassandra Mine earlier. In fact, much of *De Re Metallica* is concerned with these difficult ores and their repeated roastings. The enormous amounts of niter used to reduce these ores is given in Brown (1907): sphalerite needs 25 parts of litharge; iron pyrites, 50 parts; arsenical pyrites, 40 parts; chalcopyrite, 30 parts; and fahlerz (copper, antimony, and sulfur), 35 parts. Incidentally, borax is not needed with the quartz and fluorite gangue that have been seen with the Ag-Co group.

With all the above ingredients mixed together in a crucible, the first part of the cupellation process can continue. The crucible is put in a furnace and gently fused, a process that takes about 30 to 40 minutes. It can be done in a charcoal fire if the charcoal is heaped around the crucible. [Note that tongs would not be needed, as the crucible can be smashed when cold.] After fusion, the liquid is then poured into a preheated conical mold and left to cool (see pl. 5b). Slag forms on the top—invariably black—and a button of lead at the bottom, ideally weighing between 28 and 32 grams (pl. 5c). Sometimes there is a thin layer of matte between the button and the slag. The glass is then smashed with a hammer (pl. 5d) and the lead button easily freed from the slag. Next, the button is hammered into a cube whose corners are flattened so that they will not damage the cupel in the next stage (pl. 5d,e). The objective is to produce a cube of lead, rich in gold and silver and weighing about 28 grams.

The cupel requires careful preparation and is made today as in Agricola's time. First, animal bones (or the ashes of hardwood) are repeatedly boiled in water to remove their grease and organic matter; then they are calcined, finely ground, and washed. A mixture of 1.5 parts bones, 1 part wood ash, and .5 part of the ashes from the scrapings of hides is a good one. Ashes from the horns of deer are the best, as well as those from the skulls of animals and fish bones (rich in phosphorus). The ashes are sprinkled with beer or white of egg or cows' milk. In a 1907 recipe (Brown 1907:115; following Mitchell 1888), the bone ash is mixed, one pound at a time, with a strong solution of pearl-ash (carbonate of potash) in warm water until it sticks together. It must not be too hard, but when it is squeezed in the hand, it should show the imprint of the fingers like a snowball. The texture of the cupel is very important. If it is too compressed or made of too fine ash, the cupel will crack in the furnace. If too coarse, the cupel will absorb the silver, the last thing one wants. The bone-ash mixture is then molded into shape and dried in a warm place for several days.

Next comes the difficult third stage of the cupellation process. The cupel must be warmed before the cube of lead is placed in it (tongs required) and then put in a muffle in a furnace (fig. 2a). [Nineteenth-century assayers used three different types of furnace: for scorification, crucible fusion, and cupellation.] The muffle is a fireclay box with the front end open and a front door, also of fireclay. After the cupel is inserted, a piece of glowing charcoal is placed in the mouth and the door closed. The furnace temperature is critical—it must be between 807°C and 926°C.

After ten minutes the muffle door is removed and the lead is seen to have melted into a liquid in the cupel (lead melts at 327.4°C). The cupel must not reach white heat, nor must the lead bubble. It must simmer gently at about 875°C. If the temperature is too low, the lead solidifies into a spongy mass and will not melt again. The lead quietly volatilizes in the cupel at this important stage until the critical point is reached when the last of the lead leaves the silver bead. This stage is called "brightening," "flashing," or "blicking." The lead appears to revolve with great rapidity and rainbow colors are seen. Finally a film passes over the bead and no more action is visible. The cupellation is now completed. In figure 1b, the assayer is watching this process. Very rich ore bodies betray themselves by the peculiar mottled appearance of the molten lead shortly after the cupellation begins. The luminous blotches of litharge cover the lead like a net as they form. The cupel is now moved by degrees to the cooler front part of the muffle and covered with an inverted hot cupel, as a rich silver bead spits and is also volatile. When cool, the bead will be the size of a pea if the ore is fairly rich. With a rich galena it is about the size of a match head or peppercorn, while with a 180-ounce galena the bead is even smaller (see pl. 5b). The bead usually contains some gold except with the Ag-Co group when it contains some bismuth. Today, a gold-rich bead is parted with hot, dilute nitric acid, dissolving the silver which is later precipitated out of the solution. Doré beads containing less than 66% silver do not part. The bead is weighed and, according to the assay/ton system, if it weighs 80 milligrams, the ore will run at 80 ounces: if 20 milligrams, 20 ounces; and so on. This is the purpose of cupellation, an intricate process probably beyond the capacity of ancient man.

A good cupel absorbs its own weight in litharge, while the other 15% of the lead is volatilized. A too small cupel does not absorb all the litharge. The assayer must watch the cupellation process carefully to stop it before the silver also volatilizes away.* It is worth noting that a cupel *can be used only once.* Therefore, once again, if cupellation had been practiced, one should find many abandoned cupels and pieces of black glass. There should also be lots of glass-stained sherds if roasting was done in crucibles or scorification dishes were used.

Thus, there are three basic smelting processes:

1. Scorification or roasting in a shallow clay dish;
2. Crucible roasting;
3. Cupellation.

In 1800 B.C. native silver, found on the surface with "horn silver" ($AgCl$), was probably simply roasted with its quartz and fluorite gangue materials in a clay pot over a hot fire [as this writer did in Nevada in 1983].*

---

*Agricola shows him looking through a slit in a board, to protect his eyes and face from the heat of the furnace (Hoover 1912:223; fig. 2b).

---

*Agricola (Hoover 1912:483) gives a clue here when he states that "the Moravians and Carni, who very rarely make more than half a pound of silver, separate the lead from it [if indeed the ore contained any lead, which is doubtful. Agricola seems to have started the myth that silver can only be produced from or with lead] . . . neither in a furnace resembling an oven, nor in a crucible covered by a dome, but in a crucible which is without a cover and exposed to the wind: on this crucible they lay cakes of silver-lead [?] alloy, and over them they place dry wood, and over these again thick green wood. Then they use a bellows."

It is difficult to pinpoint the exact location of Moravia from Agricola's text, except that it was the southeast border part of Bohemia, from the Danube at Bratislava to Ratibor in Silesia via the River Morava through the Moravian Gates to the Upper Oder. He may well be referring to the rich silver deposits of Banska Stiavnica (Schemnitz), some 85 miles east-northeast of Bratislava, where the silver always contains gold. Bohemia certainly covers the area of the ancient Únêtice Culture which, by 1800 B.C., had tin bronzes, silver, and glass beads (see map 3). The geological structure is of the Hercynian Age, the belt of high-temperature rocks which stretch through Morocco, Portugal, Brittany, and Cornwall with their stannite, tin, silver, and copper ores to the Erzgebirge. Sardinia with similar "dry" silver ores and some tin is an offshoot of this structure.

4   THE ULU BURUN (KAŞ) SHIPWRECK AND OLD TRADING ROUTES

Since completing examination of the data and artifacts and the metallurgical experiments described here, this author feels that his hypothesis, that the first glass was accidentally discovered by the smelting of a certain rare type of silver ore which contained traces of the scarce metal cobalt, has been confirmed by the results of the excavation of the Ulu Burun shipwreck off the coast of Turkey near Kaş by George Bass (1986). The cargo, dated very accurately to around 1400 B.C. by the Mycenaean, Cypriot, and Syrian pottery on board, contained some 80 oxhide ingots of copper, *but also oxhide-shaped ingots of pure tin.* In connection with this book, the most important finds were some 20 or more "round glass ingots" which were colored by cobalt. Brill (1974) has confirmed that the glass is similar chemically to that of Mycenae and Egypt, whereas the glasses of Mesopotamia and Iran, or perhaps western Asia in general, are of a different chemical type (Bass 1986:282). Bass interestingly suggests that Egypt had obtained glass as a *raw material* from the Palestinian coast and that possible shipments of glass ingots are mentioned in the Amarna tablets from the King of Alasia (ibid., 293). This writer would suggest that the glass came via an entrepôt, either on Rhodes (Ialysos), Cyprus (Enkomi), or the Levantine trading ports of Minet-el-Beidha, Byblos, and Tell Abu Hawam, which have clear Mycenaean connections.

The round glass ingots were approximately 15.5 centimeters (6.2 inches) in diameter and 5.5 centimeters (2 inches) deep, with a slightly concave top and tapered towards the base; exactly the shape to which a glass slag would cool in a small, open silver smelting furnace or crucible (see fig. 9b). The glass is similar to the piece of blue cobalt slag found in Saxony by Siegfried Flach on the silver-smelting slag heaps of Schneeberg (pl. 3c) and subsequently analyzed by this writer. The fact that the 15 glass ingots listed by Bass varied so much in weight, from 2,607 grams to 1,597 grams (about six to four pounds), indicates that they were the products of different batches of ore which varied in silica (i.e., glassmaking slag) content. The diameters appear

constant and it is interesting that these "one-off" products of a smelting site were not remelted to a constant ingot size and weight but traded as raw material to the glazing centers of the Levant and Near East, as was the rest of the cargo.

The Ulu Burun shipwreck with its assemblage also throws light on the trade routes of the Late Bronze Age in Europe, the Near East, and even Asia. The wreck is full of diagnostic objects: a bronze double-axe blade; a bronze sword; a Mycenaean seal; glass, faience, and amber beads (the latter from the Baltic; Beck in Bass 1986); and faience rhyta.

Rhyta are also found at Mycenae, House of Shields (Dayton 1978, pl. 19:5) and Kition (ibid., pl. 21:2), Phylakopi, Thera, Ephesus, and Gournia (ibid., 301, 303). One of the faience rhyta was in the form of a ram's head, as seen at Enkomi (Dayton 1978:361; Murray 1900:37). Rhyta is seen in the wall paintings of the tombs of Rekhmiré (around 1435 B.C.) and of Menkheperesoneb (Dayton 1978:286) where kilted traders are depicted carrying Sherden-type swords and bearing oxhide ingots. Some of these are painted white and Bass suggested (1967) that they could have been of tin [this writer thinks they could also have been of silver; see fig. 6].

The faience bead illustrated by Bass (1986:289, KW 32) is typically Mycenaean and Levantine, as found in the Tomb of Clytemnestra, in Nauplion Chamber Tomb B, and in Alalakh III, and is a type common in the Indus civilization (Dayton 1978:206, no. 17; 426, fig. 27, no. 4; Marshall 1931:468, pl. 157:41). Large slabs of blue frit also were found in the Cape Gelidonya wreck (Bass 1967) (see pl. 8). These were a product of copper smelting, as distinct from silver/cobalt smelting. Copper blue frit was used as a pigment.

The Mycenaean world was abundant in tin bronze, tin itself (plated on clay; Higgins 1967, fig. 131), and tin foil found in the Shaft Graves (Karo 1931). Important in making the case for western sources of metals is the fact that in Sardinia a crucible has been found with about ten kilograms (22

MAP 5

The Tin Region of Sardinia

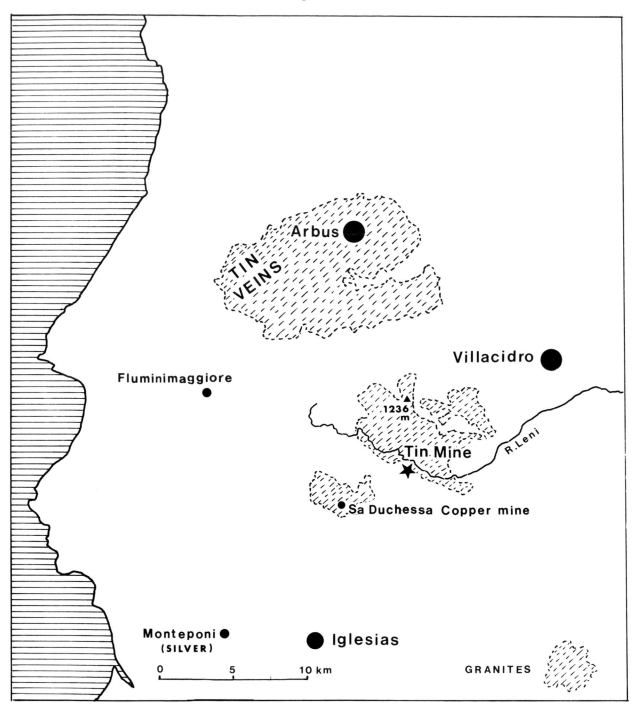

TIN VEINS

Arbus

Villacidro

Fluminimaggiore

1236 m

Tin Mine

R. Leni

Sa Duchessa Copper mine

Monteponi
(SILVER)

Iglesias

0   5   10 km

GRANITES

pounds) of cassiterite in it, while a lump of melted tin was found at Lei (Sa Maddalena) weighing 700 grams (Guido 1963:153). Tin ingots have been located in the sea off Haifa (Wachsmann and Raveh 1981:160). These finds refute the former notion that tin is unstable and disintegrates, so it cannot ever be found on archaeological sites.

The remarkable find of five hoards of bronze in the gold-mining region of the Banat, along the Mures River, which contained no less than 4,000 kilograms of bronze and 300 kilograms of pure tin ingots (which must have come from the Erzgebirge), has been overlooked by Middle Eastern archaeologists. A further 70 hoards belong to the early Hallstatt period. The hoards also contained Baltic amber and blue glass beads (Rusu 1963:184). This remarkable bronze-working industry was concentrated in the classical gold-mining area of Dacia and shows the existence of highly organized long-distance trade in tin and cobalt glass beads from Bohemia and amber from the Baltic.

It seems that Bass, encouraged by the ideas of Muhly and Maddin, dogmatically assumes that the ship was sailing "from east to west" and that the trade was in the hands of "Canaanite" merchants, whoever they may be. Numerous groups of peoples entered the Levant both by land and by sea during the Middle and Late Bronze Ages. Hyksos, Keftiu, Mycenaeans, Sherden, and finally the Philistines appeared along with other Sea Peoples. They brought with them Cyclopean masonry, bronze swords and socketed axes, torques, and, in the case of the Hyksos, camps surrounded by massive earthworks reminiscent of central and western Europe. A few hundred years later, there is clear evidence of Phoenician sea trade covering the whole of the Mediterranean, out to the Atlantic, and using ships similar to those of the Ulu Burun wreck.

Bass (1986:272) comments that "the presence of tin in the Cape Gelidonya wreck has not been accepted by Maddin, Wheeler, and Muhly . . . even though tin was identified by separate laboratories in Turkey and the United States," while again the trio query his sampling techniques (ibid., 276, note 25). As Bass mentions (ibid., note 9), Muhly discusses the Gelidonya wreck without reference to its primary publication, i.e., Bass 1967.

If tin *must* come from the East to satisfy the epigraphist [and Muhly dogmatically states that *annaku* is tin], then the tin, on geological grounds, must come from either eastern Siberia or from Burma and Malaysia via the Indus civilization, which is not an impossibility. Although great em-

phasis has been placed on tablets mentioning the transportation of tin *from* Eshnunna *to* Mari (i.e., from east to west), what Muhly and colleagues seem to have overlooked is an important quotation from Hutchinson (1950:55): "There is some significance in Sargon of Akkad's references to the Tinland beyond the Upper Sea [Mediterranean] and in the direction of Kaptara [Crete?]."

To the question of the source of tin in the Bronze Age is now added that of the source of the scarce elements cobalt, antimony, and silver. Amber beads from the Baltic are cheerfully accepted, although the amber route from the Baltic to the Adriatic passes right by the deposits in central Europe of copper, cobalt, tin, silver, bismuth, amethyst, antimony, carnelian, and blue anhydrite where it is known, by Carbon 14 dating, that the wealthy metal-working civilizations existed soon after 2000 B.C. (see map 3).

It is clear that various groups of peoples entered the Mediterranean basin and reached the Levant after the explosion in metallurgical technology some time after 2000 B.C. Some of them, e.g., in Spain, may have been indigenous. With the advent of metal tools came the building of sophisticated ships and long-distance trade by sea. Goods, particularly metals, were traded overland from village to village following the earlier routes in obsidian and jadeite. When brought to sheltered inlets, they were exchanged for desirable goods [probably slaves or textiles]. Or they were traded by ship along the coast, a method eminently suitable for the transportation of metal which is heavy. In the case of Mesopotamia, a long land route was needed to reach either Anatolia and beyond or Syria and the Levant, as seen in old Assyrian records.

Bronze was obviously the most needed commodity and along with glass trinkets and faience beads a useful bartering currency. Again, silver and gold were luxury items. The peoples of the Old Palaces of Crete obviously traded by sea, as did their successors of the New Palace period. The oxhide ingots of Haghia Triada came from Sardinia where, at the same period, there was a flourishing metallurgical industry. No glass has been found in Minoan Crete, although the Minoans had masterpieces of quite un-Egyptian faience in the form of the Town Mosaics, the Snake Goddesses, and so forth.

Contemporary with the New Palace period of Crete and Santorini was the civilization of the Shaft Graves of Mycenae which also was rich in bronze, silver, electrum, and gold, as well as much amber, faience, rock crystal, amethyst, and, *above*

FIGURE 6

Scenes from the Tomb of Rekhmiré (A–T) and Amenmose (U–V), ca. 1435 B.C.

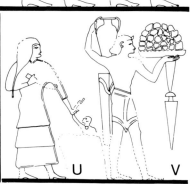

6. The reliefs depict "the arrival in peace of the chiefs of Keftiu-land and the islands which are within the Great Sea" (A–M). Davies (1943) emphasizes that Keftiu is not defined as an island but as a race and culture with connections covering Crete, Cyprus, the Helladic Islands, and the Mainland on both sides of the Aegean. Scholars, influenced by Sir Arthur Evans, have tended to forget this, and to identify the Keftiu solely with Crete.

We now know that many of the ox-hide ingots of copper came from Sardinia, together with the Sherden swords. However, the Retnu (T) also carries an ingot and a Hyksos-type bow.

The trade goods of the Keftiu include large quantities of silver—vessels, rings, and blocks (J). The silver vessels and rhytons are typically Mycenaean in style; necklaces of blue beads indicate that *kyanos* or cobalt blue beads were not produced in Egypt at this time; and there is also a dog's head of pale gold (electrum?).

Figures N to T depict "the arrival in peace of the chiefs of Retnu and all the lands of Further Asia." They wear clothing suitable for a cold climate, yet the chief (Q), armed with three swords and a quiver, is followed by his Keftiu servant bearing his Hyksos-type bow and very Mycenaean-looking plumed helmets. The Retnu also bring a chariot with 4-spoked wheels, as seen on the signet ring from Shaft Grave IV and the stele from above Shaft Grave V at Mycenae, while the horses led by another figure remind us of the famed breed from the Plain of Argos.

The Retnu also bring gold and silver rings, red and blue beads, collars of blue, red, and gold beads probably from Mycenae, jars of blue and green glass, and baskets of green and red material (perhaps for glass-making). Olive oil and unguents are other items, and women slaves are shown wearing the heavy triple skirts that we see in Crete and at Mycenae.

Two unique types of figures are represented (N and T), both archers, one with an ingot, the other with a jar of olive oil. Note also that Retnu (V) has a Keftiu kilt, a quiver, a Sherden sword, and a tray bearing lumps of ore or glass.

all, cobalt blue glass. The early Mycenaeans must have been trading, via the Adriatic, to the Polada culture of northern Italy, again by sea. There existed, also at the time of the Shaft Graves (ca. 1650 B.C.), the mysterious foreign Hyksos Dynasty in Egypt and the Levant, where in their coastal cities of Gaza, Byblos, Ras Shamra, and Alalakh there was a wealth of bronze, silver, and faience.

Earlier and contemporary with the Old Palaces of Minoan Crete and the Egyptian Twelfth Dynasty were the trading posts in early Assyria (Anatolia) dealing in the mineral wealth of the Danube basin, the Balkans, and even Bohemia. In northern Greece, the peoples with Minyan Ware were trading via the Vardar Gap. A related group, also with gray wares, traded via the Black Sea to the Caucasus, northern Iran, and beyond. It appears to the writer that the famous siege of Troy was a battle over these trade routes.

By 1400 B.C., with the decline of Minoan Crete, the Mediterranean, as far as Sardinia and Spain, was a Mycenaean sea. The Sherden appear for the first time in Egyptian records (ca. 1450 B.C.), thus bringing Sardinia onto the pages of history. They were warrior mercenaries serving in the Egyptian garrisons of the Levant where Mycenaean trade goods were abundant. It is to this time that the Ulu Burun shipwreck is dated, with its cargo of copper and tin ingots, faience beads and rhyta, cobalt blue glass disc ingots, and ivory. The ivory could well have come from Libya: the Libyans were allies of the later Sea Peoples who attacked Egypt at the end of the Late Bronze Age, when the Sherden were both defenders and invaders. Six groups are mentioned on the monuments of Ramses III at Medinet Habu, while other Egyptian records list eight more names. Obviously, this federation of peoples came from different islands and parts of the Mediterranean. The interesting question is, what was the route of the fully laden Ulu Burun ship? The pottery, the seal, and the sword imply that the crew were Mycenaean seamen from one of the Mycenaean trading posts which existed in the eastern Mediterranean.

The copper ingots could have come from Cyprus, Ergani Maden, Laurion, central Europe, Sardinia, or Spain. Lead isotope analysis will indicate their provenience. The tin ingots must have come from the Erzgebirge, Spain, or Sardinia. The cobalt glass ingots could only have come from the Erzgebirge. The ivory could have come from North Africa or from Syria, where, at this time, Thothmes III hunted elephants in the marshlands east of Aleppo. The faience could have been made on Cyprus, in Syria, on Crete, or on the Mycenaean mainland. Again, trace-element and lead isotope analysis will help.

It is possible that the ship traveled to Sardinia for copper, then to the head of the Adriatic for cobalt glass and Baltic amber, then to Libya for ivory, and then eastwards to the Levant, but more likely to some central Mediterranean entrepôt, still to be located. It carried the typical cargo brought by Keftiu, Syrians, and Mycenaeans to Egypt at exactly this date, 1400 B.C., but is well off course for Egypt, being on the southern shores of Turkey, between Rhodes and Cyprus. If its destination was westwards, it would hardly be carrying tin and copper to Mycenae, Tuscany, or Sardinia, and cobalt glass and amber to the Adriatic. Nor, if it was heading for the Greek mainland, would it carry the tin and glass to the east and then back again westwards. In this writer's opinion, the ship was clearly heading for Cyprus, hugging the shore, and then sailing to either the Levant and/or making a run down the coast of the Philistines to Egypt. Since it had a full, mixed cargo of raw materials, it was obviously heading for a manufacturing center, the Birmingham or Pittsburgh of the Levant. Cyprus, safe across the sea, seems a prime candidate. From here, manufactured goods were traded to Syria, Palestine, and Egypt. It will be interesting to see the geological origin of the copper ingots. Cyprus seems unlikely as this would be similar to sending "coals to Newcastle."

The ship's mixed cargo had apparently been assembled somewhere at a convenient trade center, where other ships called with single cargoes of copper, ivory, tin, and glass. Crete immediately springs to mind as such a center, an easy sea journey to the African mainland and equally so to Sicily, Sardinia, and Spain; to the Peloponnese and the Adriatic; and eastwards to Santorini, Rhodes, and Cyprus, where its cargo was turned into manufactured goods for onward shipment.

FIGURE 7

Lead Isotope Plot of Major Copper and Silver Deposits of the Old World

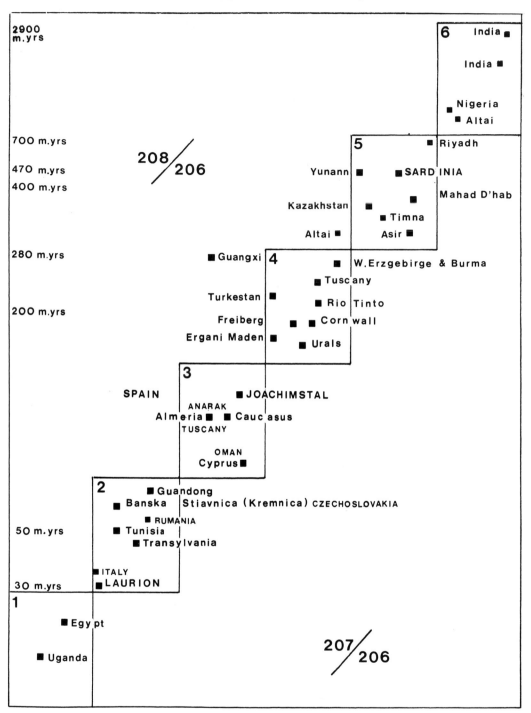

7. It can be seen that the deposits have a geological age varying from under 30 million years to over 3,000 million years as in the case of India. The deposits are formed at different times according to when the minerals are mobilized from the magma by tectonic action within the earth's crust.

5 LEAD ISOTOPE ANALYSIS

After the revolution of Carbon 14 dating, there is now an important new geological technique which can identify the source of artifacts, both metal and glass, if they contain even as little as a few parts per million of lead (see Dayton 1978; 1981a; 1981b; 1984a; 1985; 1986; Brill, Barnes, and Adams 1975; Chamberlain and Gale 1976).

As a result of new lead isotope analysis, this writer was able to show, in a paper given at Heidelberg in April 1990, that the silvers of EC II Syros and Amorgos, two silvers from the royal tombs of Ur, a silver ingot from Khafaje, silver from the Shaft Graves of Mycenae, and the important silver from Tomb 146 at Abydos came from the rich silver mines of Almeria [later worked by Hannibal with 40,000 slaves]. Coins made of Almeria silver together with an ore sample from the British Museum of Natural History provided this writer with the vital analytical control. Further, it was shown that two early second-millennium silver figurines from Syria came from the silver-rich, southwest region of Sardinia (see isotope plot, fig. 8, map 5).

Khafaje and Ur are contemporary with the time of Sargon of Akkad who, as mentioned before, "had a silver mountain over the sea and to the west." This mountain was evidently in Almeria. The vital analyses were in Gale and Gale (1988), who did not make the connection, and, for this writer, by Professor Ron Farquhar of Toronto University.

Lead isotope analysis is based on the principle propounded in 1903 by Rutherford and Soddy of the disintegration of radioactive elements, particularly uranium, thorium, potassium, rubidium, and carbon. An atom of lead contains four isotopes. The lead 204 isotope, known as primeval lead, is stable and was formed at the same time as the earth itself. Lead 206 is derived from the decay of uranium 238, with a half life of 4.4638×10^9 years; lead 207, from uranium 235, with a half life of 0.70381×10^9; and lead 208, from the decay of thorium 232, with a half life of 14.01×10^9 years. In an atom of lead, the ratio of the four isotopes varies according to the age of the lead: i.e., how long the three nonstable isotopes have been decaying. Thus, lead in an ore deposit or in a rock can be dated. For example, the lead deposits in Tunisia have been dated to 50 million years ago, in northern Nigeria to 540 million years, and in South Africa to 2,300 million years. In archaeology, the 207/206 ratio is plotted against the 208/206 ratio to give a growth curve as in figure 8.

Two factors make lead isotope analysis useful in archaeology. First, only a minute amount is needed; and second, lead is not affected by heating and smelting. A trace of lead is thus equally useful in a bronze, a silver, or a glass artifact. Obviously, the method is useless in a relatively modern time frame: during the time of the Roman Empire, scrap from many areas was remelted and mixed. However, the method is effective with related groups of artifacts, such as ingot torques, a group of artifacts from Cornwall, Cartaginian coins from southern Spain, and Athenian coins from Laurion near Athens. It can be seen in figure 8 that objects from Sardinian ore are quite distinct, as is copper from Cyprus. Of two weapons from the same site and date from Beth-Shan in the Jordan Valley, garrisoned by Sherden mercenaries around 1400 B.C., one came from the distinctive Sardinian copper, and the other from equally distinctive copper at Laurion. The lead in the yellow glass from Nimrud, about 800 B.C., came from the classic mining area of Carinthia where antimony also occurs. Lead with antimony makes a distinctive yellow glass. Even negative evidence is of value, where it is clear that the copper slag from Isili in Sardinia must have some high percentage of Sardinian copper in it for it to plot so near to the Sardinian field. Sardinian metallurgists were obviously mixing copper or scrap from other areas [probably Spain] with Sardinian copper.

A startling result of this technique showed that the copper oxhide ingots from Haghia Triada, dated to around 1600 B.C., similar to those on the Ulu Burun wreck (see pl. 8), came from unmistakable Sardinian copper (Dayton 1984). It is to Bass's credit that as long ago as 1967 he should have suggested that the many oxhide ingots found on Sardinia

FIGURE 8

Lead Isotope Plots

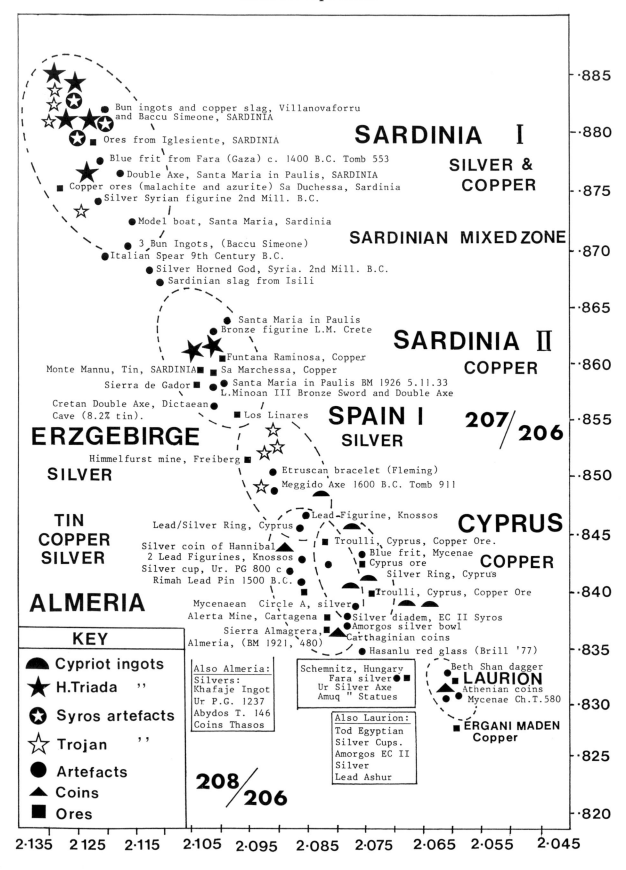

FIGURE 8

Lead Isotope Plots *continued*

Lead isotope ratios for selected artifacts (Dayton at Heidelberg 1990, unpublished) plotted against ore deposits.

The analyses clearly show that blue frit from Mycenae around 1300 B.C. came from Cyprus, Carthaginian coins from Spain, and Athenian coins from Laurion. Recent analyses by Gale and Gale (1988:135) show that the Haghia Triada oxhide ingots came from both the Funtana Raminosa and Sa Duchessa copper deposits in Sardinia. [Curiously, the Gales think that the ingots come from Cyprus (ibid.)]. Minoan Crete was certainly in contact with Sardinia. The writer's recent analyses (1990) of two silver figurines of the early second millennium from Syria show that they are made of Sardinian silver, without any doubt, as the south-west Sardinian ores are very distinct isotopically. The exciting discovery by the writer, plotting a British Museum sample from Almeria (Dayton 1978:443) is the importance of southern Spain as a source of silver from the mid-third millennium B.C. Khafaje, Ur, Egypt (tomb 146 at Abydos with its Kamares pottery), later Mycenae of the Shaft Graves and early Cycladic Syros and Amorgos, all obtained silver from this classic area where Hannibal later had 40,000 slaves working in the "Mons Argentarius" of southern Spain. Troy, however, got its silver from Sardinia *and* from the Erzgebirge. Syros also received silver from the Erzgebirge. The Carthaginian coins provided a perfect control for the British Museum sample which was analyzed for the writer in 1976 by Valerie Chamberlain at the University of Alberta, Edmonton, and also by the late Lynas Barnes at the U.S. Bureau of Standards, Washington, D.C. There are no silver deposits on Cyprus.

The fortifications of Los Millares, Zambujal, El Argar, and other sites along the Rio Almanzora have been compared by earlier archaeologists with those of Chalandiani, Mycenae, and the Levant. The map shows how these sites are all located near the rich mineral deposits of the area. Six years of excavations at El Cerro de la Mora near Grenada have yielded carbon 14 dates in an unbroken sequence from 1700 to 790 B.C. (J. Carrasco 1987). At Zambujal a smelting site was discovered by Schubart and Sangmeister, and the cyclopean walls of the settlement were found to be 15 meters thick (48 feet).

In 1839 the silver mines of Almeria were re-opened. The companies worked the old Roman and Carthaginian shafts. The La Caridad mine worked lead-free silver and dry silver ores grading 72% silver. In 1851 the area had 290 working mines and 45 smelters. In 1882 the area produced 804 tons of copper, 151 tons of tin, and 115,764 tons of argentiferous galena. Earlier in 194 B.C. the Roman general Porcius Cato took to Rome 25,000 pounds of silver in ingots, 123,000 pounds as silver coin, and 1,400 pounds of gold (J. L. Andres Sarasa 1982; F. Villasante 1915; J. Jauregui and E. Poblet 1947). There are no records of cobalt production.

The isotope data on the plot is derived from many sources which are listed in Dayton 1986. Other analyses have been carried out by Ian Swainbank, and Professor Ron Farquhar of Toronto.

FIGURE 9

The Cobalt-Nickel-Silver-Arsenide Veins of Europe

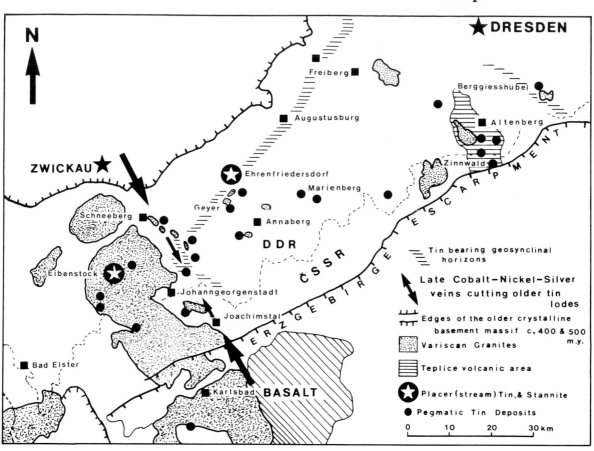

a

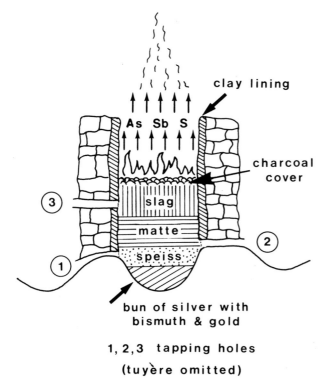

b

9a. The Erzgebirge or Ore Mountains were rich in minerals, amethyst, fluorite, rock crystal, tin, copper, and also in naturally mixed copper/tin minerals such as stannite ($Cu_2S.FeS.SnS_2$). Stannite is also known as "bell metal ore" of 27.5% tin and 29.5% copper. The Erzgebirge is the classic location for stannite, which on smelting [after roasting off the sulfur] would give a natural bronze. As Plate 7 shows, the enormous wealth of bronzes in Late Bronze Age Central Europe must have been derived from these ores.

9b. A simple silver-smelting kiln, showing the segregation of the materials at the end of the smelting process. The kiln would be destroyed at the end of each process to obtain the bun of silver. The disk of blue glass slag would be similar to those found in the Kaş shipwreck.

could have been made from the local copper-rich ores.

The spearhead from Beth-Shan mentioned above and blue frit from Tell Fara (near Gaza), both dated to about 1400 B.C., are also from Sardinian ore. Blue frit from Mycenae of the same date as Fara derives from Cypriot ore. Other analyses showed that a Middle Bronze 2 notched axe from Megiddo Tomb 911 came from Tuscan ore (Dayton 1978, fig. 278). It is certain that a silver ingot from Khafaje and silver from the Death Pit at Ur came from the silver-rich area of Almeria in southern Spain, confirming the earlier suggestions of such trade (Cary and Warmington 1963) (see fig. 8). No glass of any kind has been found in Bronze Age Spain or in western Europe of that period. Nor was it found in Bronze Age Anatolia. All this negative evidence confirms that the glass came from Schneeberg via the Adriatic.

It has recently been shown that Sardinia (suggested as the home of the Sherden) was an important center of metallurgy in the Bronze Age, trading with Spain (El Argar), Tuscany (Dayton 1984), and the Remedello and Polada cultures of adjacent Italy (Savory 1968:204). Many molds for metal casting have been found on Sardinia (Tylecote, Balmuth, and Massoli-Novelli 1984; Becker 1980). No Bronze Age glass or faience has yet been discovered on the island, however, so it appears that Sardinia can also be ruled out as a source of the cobalt blue glass.

Oxhide ingots have been found with 11 Sardinian nuraghi. Throughout the Iberian Penninsula, the Balearics, Corsica, and Sardinia are structures built of massive, undressed boulders. These were built in the early Bronze Age in western Europe and are known as nuraghi in Sardinia, where some 7,000 exist. They are up to three stories in height and are associated with smelting and metallurgy. This distinctive type of construction, known as Cyclopean masonry, is found in the important Bronze Age fortified towns of mineral-rich southern Spain, but also occurs at Mycenea, Tiryns, and Gla in mainland Greece; in Turkey at Bogazköy; and at the important entrepôt of Ras Shamara on the Syrian coast. It is also found in the Cycladic Islands, when bronze and silver appear. Distinctive features of this structure include corbelled passageways and casement walls as well as the huge stones (many weighing over a ton) employed in their construction. Later, in the full Mycenaean period (ca. 1400 B.C.), finely dressed masonry is used, but the structures, especially the corbelling and tholos tombs are the same (see Dayton 1978:466; Savory 1967, fig. 1; and Blance 1961).

Lead isotope analysis of an ingot fragment from a nuraghi at Serra Ilixi in Sardinia (by kind permission of Dr. U. Zwicker) showed that it was made of Sardinian copper from the southwest region of the island, with a mixture of other copper either from the central area or from southern Spain (Dayton 1985, fig.1; fig. 8). Coles and Harding (1979) were convinced that trade in silver, copper, and tin took place between Sardinia and southern Spain, and, now, lead isotope ratios confirm this as fact. Tin ingots were found in the Kaş wreck. Mycenaean sherds have been found in quantity on Sardinia (Cerruti 1979; Vagnetti 1980:166).

6 CONCLUSIONS

Origins of Cobalt Glass

The *only* area in Europe where cobalt glass could have been produced, certainly by accident, was in the "silver-cobalt-nickel-arsenide" belt which runs north from Joachimsthal to Ehrensfriedersdorf in Saxony (see map 3). The rich silver-cobalt ores are found outcropping near the surface and so were accessible to early man. The gangue was rock crystal with the associated fluorite and apatite which would have easily formed the cobalt blue glass. If lead, copper, iron, or other minerals had been present there would have been a useless brown/black slag. The blue slag was a valuable material—man's first plastic—and was undoubtedly traded to the Adriatic, Mycenae, and Egypt. The cobalt glass slag and the purity of the silver clinches the argument in favor of Schneeberg, aided by the circumstantial evidence of the new, raised Carbon 14 dates for Únêtice with its tin bronzes and the Baltic amber of Mycenae.

Bass's discovery (1986) of blue glass ingots in the cargo of the Mycenaean ship wrecked off Kaş in southern Turkey further confirms the origin of cobalt glass in Europe. This shipwreck has been dated to around 1400 B.C., the peak of the Mycenaean trading empire. The 20 blue discoid glass ingots found in it were seven inches in diameter and two inches thick, just the size the fairly sophisticated silver-smelting furnace would produce (see fig. 9b), and colored by cobalt. The wreck also contained amber beads, silver bracelets, 84 oxhide copper ingots, tin ingots, and lumps of a grayish-white material which proved to be 99.5% pure tin. The amber points to the Baltic and the route from the head of the Adriatic past the rich tin, copper, and silver deposits of central Europe. It seems much more probable that trade was in the hands of the Mycenaean seafarers, rather than in those of "Canaanite landlubbers."

Silver

If the cobalt glass could follow this route, so could the very pure silver of the ancient world. Before the Hyksos period about 1600 B.C., when cobalt blue glass was found at Mycenae, silver was twice as valuable as gold in Egypt (Papyrus Rhind). With the advent of Mycenaean traders, its price fell to half that of gold.

Native silver (pl. 1b) and its rich, dry-type ores did not need an elaborate smelting process. They were probably produced, as in Moravia in Agricola's time, in a clay crucible in a fire.

Cupellation was certainly not used before Roman times. At Laurion (with its lack of crucibles and cupels), free silver was removed from the galena by crushing and washing to produce a concentrate which was then roasted. The silver dissolved in the galena was left as reject ore to be treated by the French company 2,300 years later. The phosphorus found in archaeological objects was not introduced from bones but from apatite (pl. 2a,b). Preferential skimming may have been used at Laurion, similar to the Pattinson Process, to produce a silver-rich lead which was then boiled off, leaving the silver. The Romans used the same process to desilver lead. A hundred years ago galena crystals were hand-picked from ores, and sphalerite (zinc) was especially avoided.

Native silver and rich, dry (lead-free) silver ores in most areas of the world, from the Carpathians to Mexico, are found to contain high percentages of gold and some copper. The Ag-Co-Ni-As group is noteworthy in that it does not contain even a trace of gold or lead, and because of this purity (e.g., at Troy IIg and Khafaje, where pure ingots were found), the myth of cupellation arose. Lucas (1928a:313), however, gave analyses of seven Egyptian silvers from the Old Kingdom of about 2500 B.C. to after the Persian Conquest of 525 B.C., where the gold content ranges from 38.1% to 3.2%. The silver of Tutankhamen had 5.1% gold and Persian silver, 17.9%. Copper was found in six of the seven specimens. Silver is not found in Egypt, and the gold has copper but is virtually silver free. Lucas (1928a:282) also did not think silver was obtained from argentiferous galena before 500 B.C., which is probably correct. Therefore, it appears that before this date silver was only being obtained from groups of dry-silver surface ores: the Ag-Co-Ni-As group at Schneeberg, the semi-electrum gold-rich ores of Hungary and the Carpathians, and the silver of the Tertiary Belt of Almeria in southern Spain, near the ruins of Los Millares and El Argar (see map 1). The cobalt glass that appears with the first Mycenaeans and some of their silver must have come from Schneeberg in Saxony, along the trade route of the Baltic amber to the Adriatic. The evidence of the new, raised carbon-14 dates for the Únêtice culture with its tin bronzes prove that the Adriatic route was as important as the route down the Danube favored by Childe (1929), while connections via Sardinia and the Balearics to southern Spain have been completely ignored by archaeologists.

Bronze

Tin bronze was abundant in the Únêtice culture around 1800 B.C. and at Mycenae around 1650 B.C. It is suggested that tin bronzes, together with Baltic amber, certainly followed the silver route to the Adriatic, passing by Bleiberg with its blue anhydrite which appears, with silver, in Middle Kingdom Egypt. Stannite (Cu_2FeSnS_4) was abundant at Ehrensfriedersdorf. Gowland thought that this was the first ore used to produce bronze (Gowland 1912). Agricola reported that tin was washed in the rivers of the Erzgebirge at Schlackenwald, Erbisdorff, Altenberg, and Geyer (Hoover 1912:304). Ehrensfriedersdorf is famous for its amethyst and fluorite cubes (pl. 2c,d), which could have followed the same route to Mycenae, with rock crystal, and to Egypt, where only one small deposit of amethyst of poor quality exists at Gebel Nikeiba (Dayton 1978:180). Easily carved purple fluorite, labeled amethyst in many museums, traveled the same way. A carved amethyst portrait gem was found in early Shaft Grave Gamma at Mycenae (ibid., 243).

Trade Patterns

It is clear that existing concepts of Bronze Age trade, west of Sicily, must be reconsidered. Sardinia has rich deposits of native silver, much copper with a high bismuth content as seen in Cycladic bronzes (Craddock 1976:95–119), and a rich vein of tin with a modern mine which was only recently abandoned. It is curious that Lo Schiavo et al. (1985) should cast doubts on the description by Tylecote, Balmuth, and Massoli-Novelli (1984) of the tin deposits of Sardinia. The latter visited the tin mine of Monte Mannu, near Villacidro, and saw the tin

disseminated in chalcopyrite. When roasted, such an ore would have made a natural bronze. Lo Schiavo et al. (1985:318) also doubted if the cassiterite, found in another location in good-sized crystals of five to six millimeters, would have interested ancient metal workers. Balmuth and Tylecote had with them the experienced local geologist Massoli-Novelli. Tin is quite abundant in the Hercynian granites (the same formation as in Cornwall and the Erzgebirge). The author visited the area of the Arbureze in 1986 with Dr. Giorgio Padolino of the Instituto Giacimenti of Cagliari and collected excellent samples of cassiterite in white quartz (see pl. 8). The vein, as described by Tylecote and colleagues (1984), is in the valley of the Leni River, ten kilometers southwest of Villacidro, and was worked before and during the last war. On the adjacent island of Corsica, as earlier noted, there is a mixed copper and tin deposit north of Ajaccio and other copper and tin deposits worked by ancient man.

If the source of tin for the bronzes of the Ancient World remained a question since the writer's paper of 1971, then to this issue can be added the source of silver and the cobalt blue glass that appears at about the same time. Consider that, in the same area of Saxony, rich supplies of mixed tin and copper ores are found at Eibenstock and Ehrenfriedersdorf (see fig. 9), also a classic source of amethyst, purple fluorite, beautiful translucent carnelian (see Bauer 1968:513 for coral agate from Karlsbad in Saxony), and clear red jasper (traded to ancient Egypt), and not far from the Baltic amber route. It is no surprise that vast quantities of bronzes dating to around 1800–1700 B.C. should be found in the same area (fig. 6).

Summary

It would appear that what has been taken for granted for many years with regard to the movement of trade goods throughout the ancient world—namely, their origins, the direction in which they moved, and their routes—needs to be completely reexamined. Recent reexamination and analyses of some of the existing artifacts and data, a review of relevant literature, actual metallurgical experimentation, and lead isotope analysis, not to mention new archaeological finds, would indicate that many of the rich and sophisticated artifacts of the ancient Near East had their origins in the mineral deposits of Europe rather in the east itself as has been heretofore assumed.

APPENDIX

A cobalt blue bead of the Hallstatt period from St. Veit in Hungary was analyzed by microprobe for its exact elemental composition, and gave the following results:

SiO_2	75.56%
Al_2O_3	1.22
CaO	8.19
Na_2O	11.32
K_2O	0.29
FeO	0.50
CoO	0.09
MgO	0.15
TiO_2	0.02
SO_3	0.27
P_2O_5	0.14
Cl	1.04
Sb	0.14
CO_2 & Ash	0.07
	98.93%

(Mn and Ni both less than 0.01%)

First, it was clear that the bead was not the product of smelting a silver-rich lead ore, but of the smelting of the dry Ag-Co-Ni-As ores of Schneeberg. In view of the rich blue color, the amount of cobalt needed was remarkably low. The high so-dium and chlorine indicated that salt was deliberately added during the smelting. The phosphorus content was low and, probably due to apatite in the gangue, which must have been fairly pure rock crystal or quartz, as expected in the Schneeberg veins. The potassium, a low amount, was derived either from the gangue or the charcoal of the smelting process.

Analysis of the blue slag from the experimental Creede sample to which 0.125% of cobalt oxide had been added showed that nearly 50% had volatilized, leaving 0.07% in the deep blue glass, which also contained strong traces of copper and silver and a minor trace of nickel. The silver bead in the crucible also contained strong traces of copper, with minor traces of nickel, gold, and cobalt. These traces remained in the silver bead and in the blue slag after a very efficient modern crucible smelting process. The important thing was that lead was completely absent from the ore. In making glass from the blue slag, man would have had to heat the slag yet again in order to pour it into molds with more loss of trace elements. It is of interest that traces of copper are found in blue glass beads which owe their color primarily to cobalt, in both Egypt and the Levant during the Late Bronze Age.

BIBLIOGRAPHY

Agricola, G.
1530 *Bermannus, sive de re metallica, Dialogus.* Basel.
1546 *De Natura Fossilium.* Basel.
1546 *Rerum Metallicarum Interpretatio.* Basel.
1556 *De Re Metallica.* Trans. H. C. and L. H. Hoover, 1912, *The Mining Magazine,* London. Rep. 1950, Dover Publications Inc., New York.

Andres Sarasa, J. L.
1982 *Estadisticas.* Murcia, Spain.

Ardaillon, E.
1897 *Les Mines du Laurion dans l'Antiquité.* Bibliotèque des Écoles Françaises d'Athènes et de Rome, Fasc. 77, Paris.

Badham, J. P. N.
1976 "Orogenesis and Metallogenesis With Reference to the Silver-Nickel-Cobalt-Arsenide Ore Association." *Geological Association of Canada Special Paper* 14:559–571. Business and Economic Service, Toronto.

Barfield, L.
1971 *Northern Italy Before Rome.* Thames and Hudson, London.

Bass, G.
1967 *Cape Gelidonya: A Bronze Age Shipwreck.* Transactions of the American Philosophical Society, 57:8. Philadelphia.
1986 "A Bronze Age Shipwreck at Ulu Burun (Kaş): 1984 Campaign." *American Journal of Archaeology* 90:269–296.

Bauer, M.
1896 *Precious Stones,* Trans. L. J. Spencer, 1968. Dover Books, New York.

Baumann, L.
1970 "Tin Deposits of the Erzgebirge." *Transactions of the Institute of Mining and Metallurgy.* (Section B: Applied Earth Science, 79B:368.)

Baumann, L., and G. Tischendorf
1974 "The Metallogeny of Tin in the Erzgebirge," in *Metallization Associated With Acid Magmatism,* M. Stemprock, ed., 3:17–28. Geological Survey of Czechoslovakia, Prague.

Becker, M. J.
1980 "Sardinian and Mediterranean Copper Trade." *Anthropology* 4(2):99–101.

Beyschlag, F., J. H. L. Vogt, and P. Krusch
1914 *The deposits of useful minerals and rocks, their origin, form and content.* Trans. S. J. Truscott. MacMillan, London.

Bianco Peroni, V.
1970 "Die Schwerter in Italien." *Prähistorische Bronzefunde Abteilung* 4:1. Beck, Munich.

Bietti Sestieri, A. M.
1973 "The Metal Industry of Continental Italy, 13th-11th Century, and Its Aegean Connections." *Proceedings of the Prehistoric Society* 39:383–424.

Biringuccio, V.
1540 *De la Pirotechnica.* Venice. Trans. C. S. Smith and M. T. Gnudi, 1942. American Philosophical Society, New Haven.

Blance, B.
1964 "The Argaric Bronze Age in Iberia." *Revista de Guimares* 74:129–142.
1971 *Die Anfänge der Metallurgie auf der Iberischen Halbinsel.* Römisch-Germanisches Zentral Museum, Berlin.

Boegel, H.
1971 *Collector's Guide to Minerals and Gemstones.* Thames and Hudson, London.

Bray, W.
1964 "Sardinian Beakers." *Proceedings of the Prehistoric Society* 30:75–98.

Brill, R. H., I. L. Barnes, and B. Adams
1974 "Lead Isotopes in Some Ancient Egyptian Objects," in *Proceedings of the Third Solid State Conference, Cairo,* A. Bishay, ed., 9–25. Plenum Press, New York.

Brown, W. L.
1907 *Manual of Assaying: Gold, Silver, Lead, Copper.* 12th ed. E. M. Sargent and Co., Chicago.

Caley, E. R.
1962 *Analyses of Ancient Glasses, 1790–1957: A Com-*

prehensive and Critical Survey. Corning Museum of Glass, New York.

Carrasco, J.
1987   "El Cerro de la Mora." *Revista de Arquelogia,* no. 72.

Cary, M., and E. H. Warmington
1963   *The Ancient Explorers.* Penguin, Harmondsworth.

Catling, H. W.
1956   "Bronze Cut-and-Thrust Swords in the Eastern Mediterranean." *Proceedings of the Prehistoric Society* 22:102–125.
1964   *Cypriote Bronzework in the Mycenaean World.* Clarendon Press, Oxford.

Ceruti, M. L. Ferrarese
1979   "Ceramica Micenea in Sardegna." *Rivista Scienza Preistorica* 34:243–253.

Chamberlain, V., and N. Gale
1976.   "The Isotopic Composition of Lead in Greek Coins and in Galena From Greece and Turkey." *Proceedings of the 16th International Symposium on Archaeometry and Archaeological Prospection,* National Museum of Antiquities of Scotland, Edinburgh, 1976.

Cleuziou, S., and T. Berthoud
1982   "Early Tin in the Near East." *Expedition* (Fall):9–21.

Coles, J. M., and A. F. Harding
1979   *The Bronze Age in Europe.* St. Martin's Press, London.

Cowen, J. D.
1955   "Eine Einführung in die Geschichte der bronzenen Griffzungenschwerter in Süddeutschland und den angrenzenden Gebieten." *Bericht der Römisch—Germanischen Komission* 36:52–155.
1966   "The Origins of the Flange-Hilted Sword of Bronze in Continental Europe." *Proceedings of the Prehistoric Society* 32:262–312.

Craddock, P.
1976   "The Composition of the Copper Alloys Used by the Greek, Etruscan, and Roman Civilizations: 1. The Greeks Before the Archaic Period." *Journal of Archaeological Science* 3:93–113.

Dana, E. S.
1932   *A Textbook of Mineralogy.* 4th ed. W. E. Ford, ed. John Wiley and Sons, New York.

Dayton, J. E.
1971   "The Problem of Tin in the Ancient World." *World Archaeology* 3:49–70.

1978   *Minerals, Metals, Glazing, and Man.* Harrap, London.
1981a   "Cobalt, Silver, and Nickel in Late Bronze Age Glazes, Pigments and Bronzes, and the Identification of Silver Sources for the Aegean and Near East by Lead Isotope and Trace Element Analysis," in *Scientific Studies in Ancient Ceramics,* M. Hughes ed. British Museum Occasional Paper 19:129–142.
1981b   "Geological Evidence for the Discovery of Cobalt Blue Glass in Mycenaean Times as a By-Product of Silver Smelting in the Schneeberg Area of the Bohemian Erzgebirge." *Revue d'Archeometrie Supplement* 3:57–61 (Actes du XX Symposium Internationale d'Archeometrie, Paris 26–29 Mars, 1980).
1984   "Sardinia, the Sherden and Bronze Age Trade Routes." *Annali dell'Istituto Orientale di Napoli* 44:353–371.
1985   "Bronze Age Europe and the Discovery of Glass," in *Application of Science in Examination of Works of Art,.* P. A. England and L. Van Zelst, eds. Boston.
1986   "Experiments in the Smelting of Rich, Dry Silver Ores and the Reproduction of Bronze Age-Type Cobalt Blue Glass as a Slag." Paper presented at the 24th International Archaeometry Symposium, Washington, D.C., 1984.

Dayton, J. E., and A. J. Dayton
1986   "Uses and Limitations of Lead Isotopes in Archaeology," in *Proceedings of the 24th International Archaeometry Symposium,* J. S. Olin and M. J. Blackman, eds., 13–41. Washington, D.C., 1984.

Dioscorides, P.
1655   *The Greek Herbal.* Trans. J. Goodyer, R. T. Gunther, ed. 1934. Oxford. Rep. 1959, Hafner Publishing Co., New York.

Dörpfeld, W.
1902   *Troja und Ilion.* Beck and Barth, Athens.

Drury, W.
1917   *Cobalt.* Her Majesty's Stationery Office, London.

Ercker, L.
1574   *Beschreibung Allerfürnemsten Mineralischen.* Ertz und Berckwarksarten, Prague.

Evans, A.
1921   *The Palace of Minos,* vol. 1. Macmillan and Co., Ltd., London.

Gale and Gale
1988   In *Studies in Sardinian Archaeology,* M. Balmuth, ed. Vol. 2. Tufts University, Boston.

Gimbutas, M.
1965   "The Relative Chronology of Neolithic and Chalcolithic Cultures in Eastern Europe North of the Balkan Peninsula and the Black Sea," in *Chro-*

nologies in Old World Archaeology, R. W. Ehrich, ed. Chicago University Press, Chicago.

Gowland, W.
1912   *Presidential Address.* Institute of Metals, London.

Guido, M.
1963   *Sardinia.* Thames and Hudson, London.

Grosjean, R.
1966   "Recent Work in Corsica." *Antiquity* 40:190–198.
1981   *La Corse avant l'Histoire.* New ed. Klincksieck, Paris.

Haevernick, T. E.
1960   *Die Glasarmringe und Ringperlen der Mittel-und Spätlatenezeit auf dem Europäischen Festland.* Habelt, Bonn.
1974a   "Zu den Glasperlen in Slowenien." *Situla* 14–15, 61–65.
1974b   "Die Glasfunde aus den Gräbern vom Dürrnberg," in *Der Durrnberg bei Hallein, II,* F. Moosleitner et al., eds. C. H. Beck, Munich.

Hammond, N. G. L.
1967   "Tumulus Burial in Albania." *Annual of the British School at Athens* 62:77–105.

Harding, A., and S. Warren
1973   "Early Bronze Age Faience Beads from Central Europe." *Antiquity* 47:64–66.

Harrison, R. J.
1980   *The Beaker Folk.* Thames and Hudson, London.

Hencken, H.
1932   *The Archaeology of Cornwall and Scilly.* Methuen, London.
1978   *The Iron Age Cemetery of Magdalenska Gora in Slovenia, Mecklenburg Collection, Part II.* American School of Prehistoric Research, Bulletin 32. Peabody Museum, Harvard University, Cambridge.

Higgins, R.
1967   *Minoan and Mycenaean Art.* Thames and Hudson, London.

Hofman, H. O.
1918   *Metallurgy of Lead.* Scientific Publishing Co., New York.

Hoover, H. C., and L. H. Hoover, trans.
1912   *Georgius Agricola: De Re Metallica.* Rep. 1950, Dover Publications Inc., New York.

Hutchinson, R. W.
1950   "Battle Axes in the Aegean." *Proceedings of the Prehistoric Society* 16:52–64.

Jauregui, J., and E. Poblet
1947   *Mineria antiqua en Cabo de Palos.* C.A.S.E. III, Murcia, Spain.

Junghans, S., E. Sangmeister, and M. Schröder
1960–   *Studien zu den Anfängen der Metallurgie.* Mann,
1974   Berlin.

Karo, G.
1930   *Die Schachtgräber von Mykenai.* Bruckmann, Munich.

Kohl, P.
1989   "The Use and Abuse of World Systems Theory: The Case of the 'Pristine' West Asian State," in *Archaeological Thought in America,* C. C. Lamberg-Karlovsky, ed. Cambridge University Press, New York.

Lo Schiavo F., R. Maddin, J. D. Muhly, and T. Stech
1985   "Preliminary Research on Ancient Metallurgy in Sardinia: 1984." *American Journal of Archaeology,* 2d series, 89:316–318.

Lucas, A.
1928a   *Ancient Egyptian Materials and Industries.* 4th ed., rev. 1962, J. R. Harris. E. Arnold and Co., London.
1928b   "Silver in Ancient Times," *Journal of Egyptian Archaeology* 14:313–319.

Mantell, C. L.
1949   *Tin: Its Mining, Production, Technology, and Applications.* American Chemical Society Monograph Series No. 51. Reinhold Publishing Corp., New York.

Marshall, J.
1931   *Mohenjo Daro and the Indus Civilization, II.* Arthur Probsthain, London.

Mitchell, J.
1888   *A Manual of Practical Assaying.* 6th ed., W. Crookes, ed. Longmans and Co., London.

Moore, J. McM.
1972   "Supergene Mineral Deposits and Physiographic Development in S. W. Sardinia, Italy." *Transactions of the Institute of Mining and Metallurgy.* (Section B: Applied Earth Science, 81B:59–66.)

Mozsolics, A.
1967   *Bronzefunde des Karpatenbeckens.* Akademiai Kiado, Budapest.

Murray, A. S., A. H. Smith, and H. B. Walters
1900   *Excavations in Cyprus.* British Museum, Department of Greek and Roman Antiquities, London.

Mylonas, G.
1973   *O Taphikos Kyklos B toi Mykinoi.* 2 vols. Archaeological Society of Athens, Athens.

Otto, H., and W. Witter
1952   *Handbuch der ältesten vorgeschichtlichen Metallurgie in Mitteleuropa.* Barth, Leipzig.

Parr, P. J., G. L. Harding, and J. E. Dayton
1972   "Preliminary Survey in N. W. Arabia, 1968." *Bulletin of the Institute of Archaeology,* University of London, no. 10:23–61.

Pearl, R. M.
1966   *An Introduction to the Mineral Kingdom.* New ed., edited and adapted by J. F. Kirkaldy. Blandford Press, London.

Petrie, W. M. F.
1931   *Ancient Gaza I.* Publications of the Egyptian Research Account and the British School of Archaeology in Egypt, vol. 53. Quaritch, for British School of Archaeology in Egypt, London.
1934   *Ancient Gaza IV.* Publications of the Egyptian Research Account and the British School of Archaeology in Egypt, vol. 56. Quaritch, for British School of Archaeology in Egypt, London.

Pliny
        The Natural History of Pliny. 6 vols. Trans. J. Bostock and H. C. Riley, 1855–1857. H. G. Bohn, London.

Pulak, C., and D. A. Frey
1985   "The Search for a Bronze Age Shipwreck." *Archaeology* 38(4):18–24.

Renfrew, C.
1972   *The Emergence of Civilization: The Cyclades and the Aegean in the Third Millennium* B.C. Methuen, London.

Riederer, J.
1974   "Recently Identified Egyptian Pigments." *Archaeometry* 16:102–9.

Riem, H.
1974   "Die spätbronzezeitlichen Griffplatten-, Dorn-, und Griffangelswerter in Ostfrankreich." *Prähistorische Bronzefunde Abteilung* 4:3. Beck, Munich.

Rusu, M.
1963   "Die Verbreitung der Bronzehorte in Transylvanian vom Ende der Bronzezeit bis die mittlere Hallstattzeit." *Dacia* n. s. 7:177–210.

Saggs, H. W. F.
1962   *The Greatness That Was Babylon.* Sidgwick and Jackson, London.

Sandars, N. K.
1963   "Later Aegean Bronze Swords." *American Journal of Archaeology* 67:140.
1978   *The Sea Peoples: Warriors of the Ancient Mediterranean 1250–1150* B.C. Thames and Hudson, London.

Savory, H. N.
1968   *Spain and Portugal: The Prehistory of the Iberian Peninsula.* Thames and Hudson, London.

Sayre, E. V.
1975   "Analytical Studies of Ancient Egyptian Glass," in *Proceedings of the Third Solid State Conference, Cairo,* A. Bishay, ed. 47–70. Plenum Press, New York.

Schliemann, H.
1878   *Mycenae.* John Murray, London.

Shore, A. F., and M. Bimson
1966   "An Egyptian Model Coffin in Glass." *British Museum Quarterly* 30:105–108.

Sieber, S.
1954   *Zur Geschichte der erzgebirgischen Bergbaues.* Halle, Germany.

Siret, L.
1913   *Questions de Chronologie et d'Ethnographie Iberiques.* Guethner, Paris.

Siret, L., and H. Siret
1887   *Les Premiers Ages du Metal dans le Sud-Est de l'Espagne.* Antwerp.

Snodgrass, A. M.
1974   "Metal-work as Evidence of Immigration in the Late Bronze Age," in *Bronze Age Migrations in the Aegean: archaeological and linguistic problems in Greek prehistory,* R. Crossland and A. Birchall, eds. Proceedings of the First International Colloquium on Aegean Prehistory, Sheffield, March 24–26, 1970. Duckworth, London.

Stanton, R. L.
1972   *Ore petrology.* McGraw Hill, New York.

Stemprok, M., ed.
1974   *Metallization Associated With Acid Magmatism.* Geological Survey of Czechoslovakia, Vol. 3, Prague.

Stone, J. F. S., and L. C. Thomas
1956   "The Use and Distribution of Faience in the Ancient East and Prehistoric Europe." *Proceedings of the Prehistoric Society* 22:37–84.

Strabo
        The Geography of Strabo. Trans. H. L. Jones.

8 vols., 1917–1932. Loeb Classical Library, Heinemann, London.

Theophilus
Libri III De Diversis Artibus. Trans. R. Hendrie, 1847, as *(An Essay) Upon Various Arts, in Three Books, By Theophilus, also Called Rugerus, Priest, and Monk.* John Murray, London.
De Diversis Artibus: The Various Arts. Trans. C. R. Dodwell, 1961. John Murray, London.

Trump, D. H.
1966 *Central and Southern Italy Before Rome.* Thames and Hudson, London.
1980 *The Prehistory of the Mediterranean.* Penguin, Harmondsworth.

Turnham, J.
1873 "Ancient British Barrows." *Archaeologia* 43:285–544.

Tylecote, R., M. Balmuth, and R. Massoli-Novelli
1984 "Copper and Bronze Metallurgy in Sardinia," in *Studies in Sardinian Archaeology,* M. Balmuth and R. J. Rowland, Jr., eds. University of Michigan Press, Ann Arbor.

Vagnetti, L.
1980 "Mycenaean Imports in Central Italy," in *Mycenaeans in Early Latium* Appendix 2 (Incunab-

ula Graeca 75), E. Peruzzi, ed. Edizioni dell'Ateneo and Bizzarri, Rome.

Villasante, F.
1915 *Memorias sobre la mineria en Murcia.* Madrid.

Wachsmann, S., and K. Raveh
1981 "An Underwater Salvage Excavation Near Kibbutz ha-Hotrim, Israel." *International Journal of Nautical Archaeology* 10:160.

Wells, P. S.
1981 *The Emergence of an Iron Age Economy, Mecklenburg Collection, Part III.* American School of Prehistoric Research, Bulletin 33. Peabody Museum, Harvard University, Cambridge.

Whitehouse, D., and R. Whitehouse
1975 *Archaeological Atlas of the World.* Thames and Hudson, London.

Yener, K., and H. Uzbal
1987 "Tin in the Turkish Taurus Mountains." *Antiquity* 61(232):220–27.

Zwicker, U., P. Virdis, and M. Ceruti
1980 "Investigations of Copper Ore, Prehistoric Copper Slag and Copper Ingots from Sardinia," in *Scientific Studies in Early Mining and Extractive Metallurgy,* P. Craddock, ed. British Museum Occasional Paper 20. London.

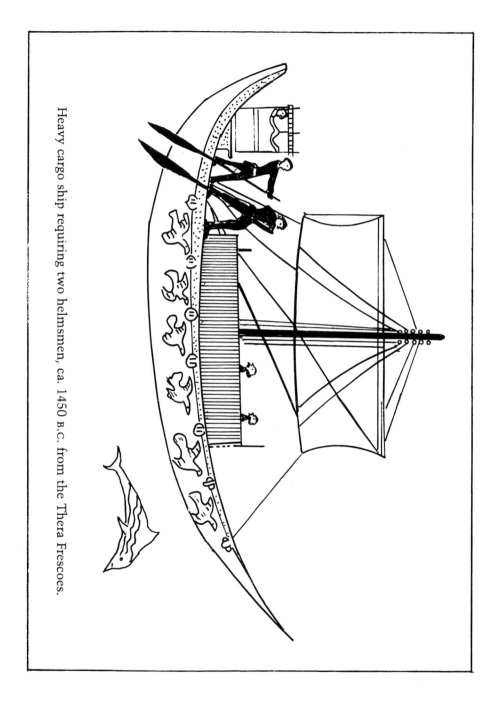

Heavy cargo ship requiring two helmsmen, ca. 1450 B.C. from the Thera Frescoes.

PLATE 1

The Discovery of Glass

a

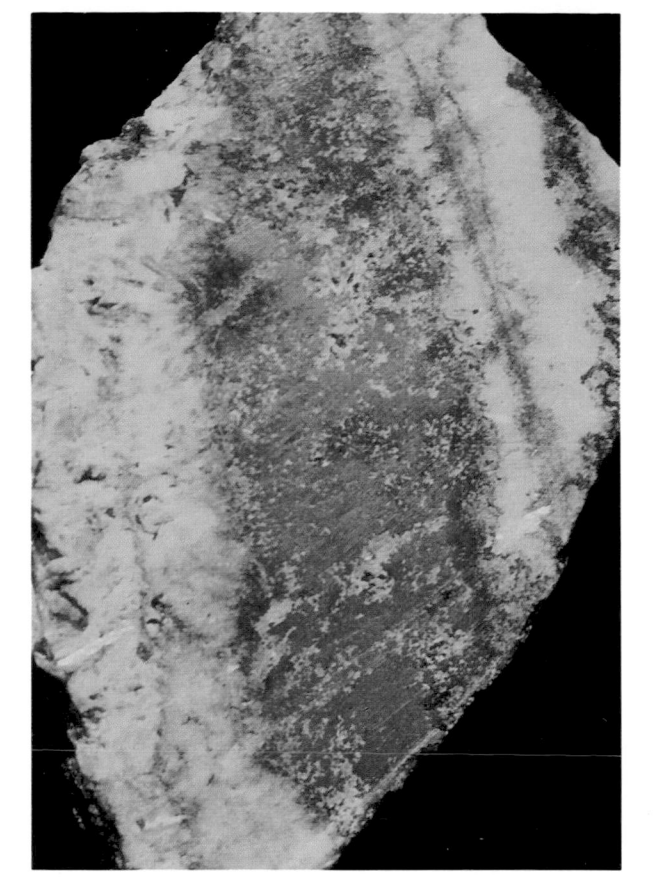

b

c

1a. The worked-out Mizpah vein at Tonopah, Nevada, which outcropped almost pure silver on the surface and also at depth. These were the ores exploited by early man, and not argentiferous galena.

1b. A typical vein of native silver in quartz and calcite with traces of cobalt. These silver-cobalt ores in quartz produced a blue glassy slag when smelted.

1c. Pink crystals of erythrite (cobalt bloom) on silver-nickel-bismuth-cobalt ore from the classic silver and color-producing location of Schneeberg in Saxony.

PLATE 2

Fluxes or "Stones that melt easily in the fire"

a

b

c

e

d

2a. A crystal of apatite (CaCl)Ca$_4$(PO$_4$)$_3$, a phosphate mineral common in the silvers of the Erzgebirge. Phosphorus is a glassformer.

2b. Crystals of green fluorite associated with the silver ores.

2c. Silver ore (cubic) with associated fluorite on quartz and calcite, from Saxony.

2d. A purple crystal of fluorite, easily carved and mistaken for amethyst.

2e. Native silver and rich silver ores occurring on a pure quartz matrix which would produce a clear glass slag when smelted.

PLATE 3

Different Colored Slags

a

b

c

e

d

3a. Green glass from the silver ores of Batopilas, Mexico.
3b. Green glass slag with traces of silver still adhering to it.
3c. Antique copper smelting slag; note the unattractive brown color. Modern slags are black and glassy.
3d. Red glass from Batopilas. The colors depend on the associated minerals in the vein as mining proceeds.
3e. A piece of blue cobalt slag, found by Herr Siegfried Flach at the site of "Agricola's hutte" in Saxony—silent testimony of a once great industry and the home of Leonardo Da Vinci's blue pigments.

PLATE 4

Experimental Smelting of Dry Silver Ores

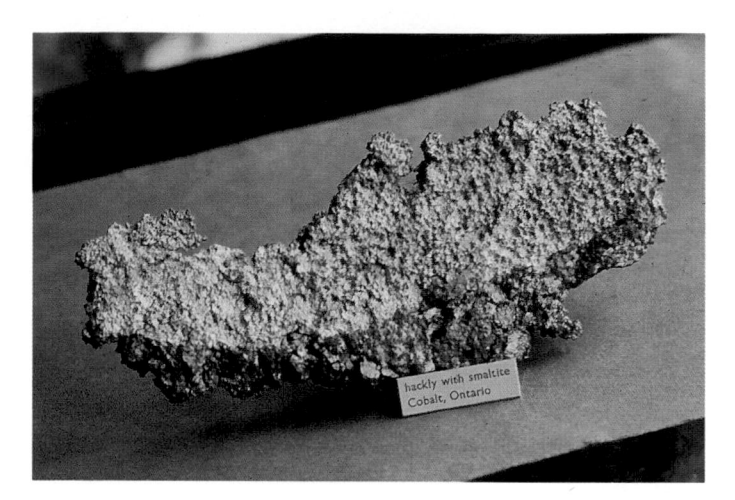

a

b

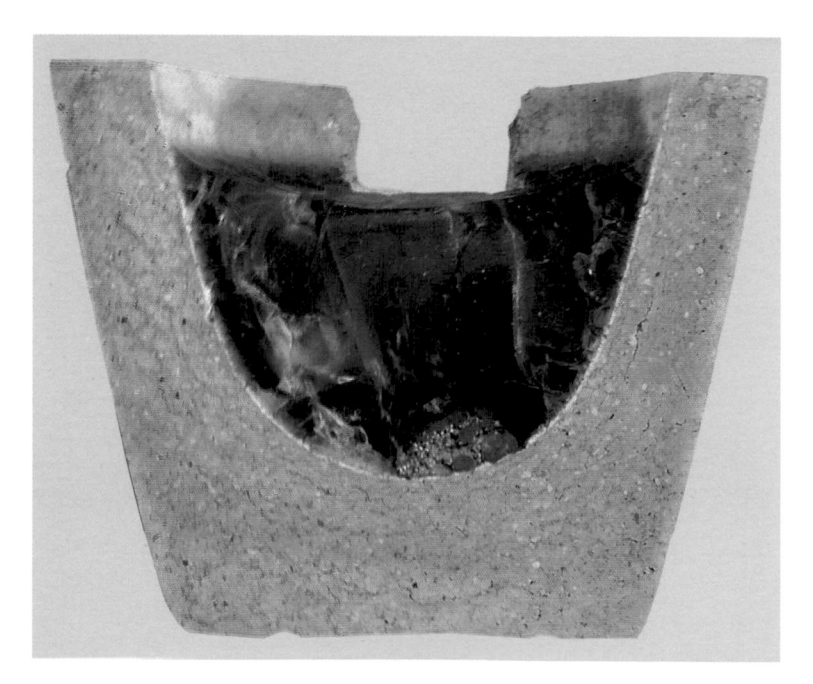

c

d

4a. Native silver (99.9% pure) on smaltite [Ag, Co, Ni, As$_2$] from Cobalt, Ontario.

4b. A fused crucible after pouring off the slag and silver.

4c. Section of the cut, unpoured crucible showing the bead of silver at the bottom, and the cobalt blue glass slag above it.

4d. Buttons of cobalt blue glass poured from the crucible in b above. The whole smelting process was remarkably simple with this type of ore.

PLATE 5

The Cupellation Myth

a

b

c

d

Cupellation is an assaying technique of considerable complexity. It could not have been used by early man, as archaeologists have assumed.

5a. The assayer with his scales, crucible, and boxes of litharge, silica, borax, sodium bicarbonate, fluorite, and wheat flour. The mix is varied according to the nature of the ore, a process requiring experience and skill.

5b. The molten charge is poured into iron molds.

5c. When cool, a cone is formed, the bottom point of which is lead plus any silver, gold, or bismuth; the upper part is the glassy slag.

5d. The lead cone is hammered into a cube and the slag discarded.

5e. The cube of lead (carefully weighed) is placed in a new, white, bone-ash cupel. The cupel is then fired in a muffle furnace: 15% of the lead boils off and 85% is absorbed by the cupel, leaving a doré bead if any gold or silver is present. The weight of this bead gives the richness of the ore. The lead is then "parted" with acid to determine the ratio of gold to silver.

e

PLATE 6

The Curious Abundance of Cobalt Glass

a

b

c

e

d

6a. A life-size cobalt glass sword hilt from Mycenae with traces of gold leaf still attached to it (Athens Museum).

6b. Very dark cobalt blue glass plaques from Sparta, around 1400 B.C., covered with gold leaf as "fake" jewelry (Athens Museum).

6c. Phoenician bottle almost entirely of cobalt glass with decoration as in 6d except the green is lead/antimony with iron. (Rhodes Museum).

6d. Phoenician head of almost black cobalt glass with lead/antimony yellow glass, and white antimony glass (University College, London).

6e. Cobalt blue glass bangles from the late Hallstatt/La Tène cultures of central Europe (Munich Museum).

PLATE 7

Central Europe and Bronze Metallurgy

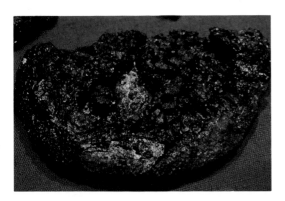

a

b

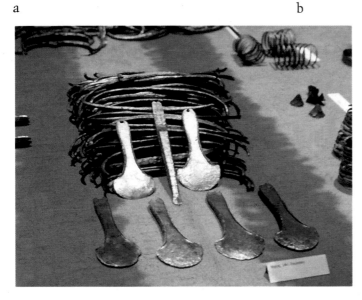

c

d

The metallic wealth of Bronze Age Europe was enormous. The objects on this page are all from the Prehistoric Museum in Munich, and include (a) raw, crude smelted bun ingots of copper; hundreds of ingot torques (b,c); together with axe blades (c). Fine molds for casting palstaves (d) were also found. These molds were of carefully chiselled metal showing the high technology of the area. Munich was an important crossing point on the route from the Erzgebirge to Italy via the Brenner Pass or the easier Lubliana Gap. Junghans (1968: pl. 29,835) found that a bun ingot of the Bodrogkerestur culture, ca. 2000 B.C., contained over 10% of tin, while ingot torgues often contained 12% or more tin. The tin must have come from the Erzgebirge.

PLATE 8

Archaeological Evidence

a

b

c

d

8a. Silver-rich Mycenae: the death mask in early grave Gamma in Circle B, ca. 1650 B.C. The Shaft Graves with their masses of tin bronzes, gold, electrum, silver, amber, cobalt-blue glass and faience are the first manifestation of the metallurgical revolution in Central Europe (Athens Museum).

8b. Ox-hide ingot of blue frit from Nimrud ca. 700 B.C., 38 cm long x 15 cm wide x 9 cm thick. An example of the trade in raw materials, such as seen in the earlier Kaş and Gelidonya wrecks (Dayton 1978:29). (Institute of Archaeology, London).

8c. The author holds a cobalt blue glass ingot similar to those found in the Kaş shipwreck. It was reproduced by him in a smelting experiment to verify the process and his hypothesis.

8d. Hallstatt/La Tène glass beads—the yellow glass is colored by lead and antimony oxides, the white glass is antimony oxide alone, while the blue glass is colored by cobalt oxide (Prehistoric Museum, Munich).